CW01334212

# PANZER III & ITS VARIANTS

The Spielberger German Armor
& Military Vehicles Series
Vol.III

Walter J. Spielberger

# Panzer III

## & Its Variants

Schiffer Military/Aviation History
Atglen, PA

Scale drawings: Hilary L. Doyle
Color illustrations: Uwe Feist

Photo sources: Federal Archives/Military Archives (3), P. Chamberlain Collection (3), Daimler-Benz AG Archives (1), Hilary Doyle Collection (2), Uwe Feist Archives (55), Robert J. Icks Collection (31), Ingo Kasten Archives (1), MAN Archives (1), Austrian Saurer Works Archives (1), W.J. Spielberger Archives (51), F. Wiener Collection (10).

Translated from the German by EDWARD FORCE.

Copyright © 1993 by Schiffer Publishing Ltd.
Library of Congress Catalog Number: 92-62386.

All rights reserved. No part of this work may be reproduced or used in any forms or by any means — graphic, electronic or mechanical, including photocopying or information storage and retrieval systems — without written permission from the copyright holder.

Printed in the United States of America.
ISBN: 0-88740-448-0

This title was originally published under the title,
*Der Panzerkampfwagen III und seine Abarten,*
by Motorbuch Verlag, Stuttgart.

---

Published by Schiffer Publishing, Ltd.
77 Lower Valley Road
Atglen, PA 19310
Please write for a free catalog.
This book may be purchased from the publisher.
Please include $2.95 postage.
Try your bookstore first.

We are interested in hearing from authors
with book ideas on related subjects.

# Contents

**Foreword**     7

**The Panzer III & Its Variants**

| | |
|---|---|
| Zugführerwagen — Development Contract, MAN, Daimler-Benz, Rheinmetall, Krupp | 8 |
| PKW III (37mm) Ausf. A, 1/ZW, Daimler-Benz | 12 |
| PKW III (37mm) Ausf. B, 2/ZW, Daimler-Benz | 14 |
| PKW III (37mm) Ausf. C, 3a/ZW, Daimler-Benz | 14 |
| PKW III (37mm) Ausf. D, 3b/ZW, Daimler-Benz | 14 |
| Panzerbefehlswagen III Ausf. D1, ex-Ausf. A, 3c/ZW, Daimler-Benz | 16 |
| PKW III (37mm) Ausf. E, 4/ZW-ZW 38, Daimler-Benz | 17 |
| Panzerbefehlswagen III Ausf. E, ex-Ausf. B, 4z/ZW, Daimler-Benz | 26 |
| PKW III (37mm) Ausf. F, 5/ZW, various | 27 |
| PKW III (37mm) Ausf. G, 6/ZW, various | 27 |
| PKW III (50mm) Ausf. H, 7/ZW, various | 27 |
| Panzerbefehlswagen III Ausf. H, ex-Ausf. C, 7a/ZW, Daimler-Benz | 36 |
| PKW III (Tp), ZW, various | 37 |
| PKW III (Amphibian), ZW, various | 39 |
| PKW III (50mm L/42) Ausf. J, 8/ZW, various | 49 |
| PKW III (50mm L/60) Ausf. J, 8/ZW, various | 53 |
| Panzerbefehlswagen III Ausf. K, 8a/ZW, Daimler-Benz | 56 |
| PKW III (50mm L/60) Ausf. L, 9/ZW, various | 56 |
| PKW III with 0725 Weapon — Experimental, ZW, Daimler-Benz | 58 |
| PKW III with Panzer IV Turret — Proposal, ZW, Daimler-Benz | 58 |
| PKW III with Pak 38, ZW, Daimler-Benz | 59 |
| PKW III (50mm L/60) Ausf. M, 10/ZW, various | 59 |
| PKW III (Fl), 10/ZW, MIAG/Wegmann | 59 |
| PKW III (75mm) Ausf. N, 11/ZW, various | 62 |
| Panzerbeobachtungswagen III (unarmed), ZW, Daimler-Benz | 68 |
| Panzerbeobachtungswagen III (armed), ZW, Daimler-Benz | 72 |
| PKW VK, 2001 (DB), ZW 40, Daimler-Benz | 74 |
| 37mm Pak (S-P) on Pz II Chassis — Prototype, LaS 100, MAN | 79 |
| 37mm Pak (S-P) on Pz IV Chassis — Prototype, BW, Krupp | 79 |
| Selbstfahrlafette III b, 2/ZW, Daimler-Benz | 85 |
| S-P for 75mm Sturmgeschütz, O series, 5/ZW, Daimler-Benz | 88 |
| S-P for 75mm Sturmgeschütz Ausf. A, 5/ZW, Daimler-Benz/ALKETT | 88 |
| S-P for 75mm Sturmgeschütz Ausf. B, 7/ZW, Daimler-Benz/ALKETT | 90 |
| 75mm Sturmgeschütz Ausf.s C, D, E, ZW, ALKETT | 90 |
| 75mm Sturmgeschütz L/33, ZW, Daimler-Benz | 99 |
| 75mm Sturmgeschütz L/43 — Experimental, ZW, Daimler-Benz | 99 |
| 75mm Sturmgeschütz 40 Ausf. F, ZW, various | 100 |
| New Sturmgeschütz — Design, ALKETT | 103 |
| 75mm Sturmgeschütz 40 Ausf. F/8, ZW, various | 103 |
| 105mm Sturmhaubitze 42 Ausf. F, ZW, various | 103 |
| 75mm Sturmgeschütz 40 Ausf. G, ZW, various | 105 |
| 105mm Sturmhaubitze 42 Ausf. G, ZW, various | 105 |

| | |
|---|---|
| 150mm sIG on Sturmgeschütz 40 — Proposal, ZW, ALKETT | 123 |
| Sturminfanteriegeschütz 33, ZW, ALKETT | 123 |
| Schnelles, Leichte Panzerfahrzeug, heavily armed — Project, Daimler-Benz | 123 |
| PKW III for Railroad Transport — Prototype, ZW, Saurer | 124 |
| Minenräumpanzer III — Prototype, ZW, Daimler-Benz | 124 |
| Pionierpanzerwagen III, ZW, Daimler-Benz | 124 |
| Schlepper III — Rebuilt, ZW, Various | 124 |
| Munitionspanzer III, ZW, various | 124 |
| Instandsetzungskraftwagen III (repair), ZW, various | 124 |
| Sturmgeschütz 40 as Ammunition Tank, ZW, various | 128 |
| Bergepanzer III, ZW, various | 128 |
| Street-Clearing Tank III, ZW, various | 128 |
| Gerät 814/Geschützwagen 634/6 for leFH 18/40/6/ Sf —Project, ZW | 128 |
| Gerät 815/Geschützwagen 634/4 (K8F 43/2 Sf) — Project, ZW | 128 |
| Gerät 816/Geschützwagen 634/3 (21 cm GrW Sf) — Project, ZW | 128 |
| Sturmgeschütz 40 with Flamethrower — Project, ZW, ALKETT | 128 |
| SU 762 — Rebuilt Russian 7.62 cm Pak (r) on PKW III, ZW | 128 |

## PKW III/IV and Its Variants

| | |
|---|---|
| PKW III n.A. (new version) — Proposal | 133 |
| PKW IV n.A. (new version) — Proposal | 133 |
| PKW III + IV — Proposal | 136 |
| PKW 604/9, ex-Panzer II long (A) 75mm Pak 43, Krupp-Gruson | 136 |
| PKW 604/10 Diesel Basic Vehicle, Krupp-Gruson | 136 |
| Fahrzeug 604/11, ex-21-559 (604-10) w/ 75mm Pak 43, ALKETT | 136 |
| PKW 604/15, Device 561 | 136 |
| Gerät 804, Geschützwagen III/IV for leFH 18/40/2 Sf | 136 |
| Gerät 807, Geschützwagen III/IV for sFH 18/1 Sf | 136 |
| Gerät 812, Geschützwagen III/IV for sFH 18/5 Sf | 136 |
| Gerät 821, 75mm L/48 Panzerjäger III/IV on Panzer IV chassis, ALKETT | 136 |
| Gerät 824, Einheitsfahrzeug III/IV as Tank, Krupp-Gruson | 136 |
| Gerät 565, Lastenträger 604/14 for 30.5 cm Mortar | 136 |
| Leichter Panzerjäger III/IV — Project, ALKETT/MIAG | 136 |
| 75mm Sturmgeschütz III/IV L/70 — Project, various | 136 |
| 105mm Sturmhaubitze III/IV — Project, ALKETT | 136 |
| Schwere Panzerhaubitze III/IV, Steel Industry | 136 |
| Munitionsfahrzeug for Schwere Panzerhaubitze III/IV, Steel Industry | 136 |
| Leichte Panzerhaubitze III/IV, Steel Industry | 136 |
| Munitionsfahrzeug for Leichte Panzerhaubitze III/IV, Steel Industry | 136 |
| Sturmpanzer III/IV | 136 |
| Flakpanzer III/IV "Kugelblitz" | 136 |

| | |
|---|---|
| **Technical Data** | **140** |
| **Bibliography** | **166** |
| **Abbreviations** | **166** |

# Foreword

The development of German armored vehicles created more or less clear prerequisites, even toward the end of the 1920s, for the composition of the eventual battle tank of the German army. Heinz Guderian's concept of a purely offensive weapon resulted in a rational division of all offensive tasks with a tank crew. The solution that was derived has served as a guide to the present day, even though technological progress has now made certain tasks unnecessary. The equipping of tanks with means of communication, which he stressed particularly, gave the German armored troops a tactical advantage even when more modern enemy tanks, better suited to the conditions, appeared on the scene. This book portrays for the reader the technical-tactical development that created a battle tank that paved the way for an entire worldwide production of tanks. The "Panzerkampfwagen III" was planned from the beginning as the standard vehicle of the German armored troops, and proved itself within the framework of the potential it was given. In determining the tactical concept, it was decided during early development to use the vehicle in the Central European area, a fact that later, in service in areas of Eastern Europe, caused insuperable difficulties. In terms of certain physical limitations, such as bridge load limits and road conditions, its designers had been misguided and thus had planned these vehicles in ways that later made it difficult to remain equal to the enemy by strengthening their armament and armor. In addition, the capacity of the tank-building industry in Germany was too meager from the start. Sufficient quantities of spare parts could be produced only in the last years of the war, by ruthlessly exploiting all existing possibilities at a point in time when bombing raids and raw-material shortages were already playing a dominant role.

The change in warfare from offense to defense was reflected above all in the tactical layout of the German armored vehicles. Here again it was the "Panzer III" chassis that remained in production until the war ended and provided outstanding service for the troops, particularly as carriers of assault guns.

In the last few years, numerous publications have appeared that have dealt in words and pictures with the action of German tanks, and particularly of the "Panzerkampfwagen III." The purpose of our book shall be to provide a technical foundation for these books. In the end it was the technicians who made the achievements of the troops possible. Along with the fighting men of all service arms, performance was attained that are worthy of being recorded historically. We hope to have produced a small contribution to these records.

As usual, the circumstances involved necessitate that a publication of this type can only be an incomplete attempt to include all the details. Too much material was lost or still remains in inaccessible places. Nor can it be the task of one individual to achieve such a compilation. Thus I am obligated once again to thank a goodly number of friends whose interest and patience contributed to making this book possible. I think above all of Peter Chamberlain, Hilary Doyle, Phil Dyer, Uwe Feist, Dick Hunnicutt, Heinrich Scultetus, K. Sarazzin and many other collaborators. The rebuilding of my archives after the war ended would not have been possible without the great understanding and generosity of Col. Robert J. Icks and Dr. Fritz Wiener. This work group will gladly accept criticism and suggestions that, we hope, will be sent to us by our readers.

Walter J. Spielberger

# Panzerkampfwagen III & Its Variants

The still-open questions about the armored equipment for the planned expansion of the Reichsheer to 63 divisions were answered to a great extent in the Army Weapons Office on January 11, 1934.

Generaloberst Heinz Guderian had, from the beginning, promoted two types of battle tank as the ultimate equipment of the armored troops. One of the vehicles was to be equipped with an armor-piercing weapon, while the other was to have a large-caliber gun and be used as a support vehicle. Both tanks were to be armed with a turret machine gun, plus a bow machine gun, in addition to their tank gun.

The vehicle with the armor-piercing weapons (the later "Panzerkampfwagen III") was to be used by the three light companies of an armored unit, while the support vehicle (the later "Panzerkampfwagen IV") was meant for the fourth company.

The work group centered around Generaloberst Guderian also had the task of determining the importance of the demands that would be made on new tanks in terms of tactical use and arranging them in the right order. These involved practical firepower, mobility and armor protection, and this division has remained valid to the present. Equally far-seeing was the five-man crew decided on then, as well as the related practical division of all tasks in the tank among the commander, gunner, loader, radio-operator and driver. The large-scale equipment of the German tanks with radio apparatus also proved to be of decisive importance later, in the war. Generaloberst Guderian, who had come from the communication troops, had advocated from the beginning that every single vehicle should have a radio and an on-board communication system, and had defended his proposal tenaciously until it was authorized.

To illustrate the difficulties under which all of this preliminary work had to be accomplished, it must be mentioned that all these plans could succeed only on paper, and that hardly any facilities for experimentation or practical testing existed.

In technical terms, German industry was able to gain experience during the development and production of the Panzer I and II, which were intended mainly for use in training, that made the development of its own ideas possible. Thus it was no longer necessary to draw on foreign developments to the same extent, even though the German developments, some of them in part very complicated, did not always take the accompanying production, supply and maintenance difficulties into consideration.

The Weapons Office issued the first development contracts for the "medium tractor" or "platoon leader's vehicle (ZW), which was the camouflaged designation for the later Panzer III, before the end of 1934. The firms involved were MAN of Nuremberg, Daimler-Benz AG of Berlin-Marienfelde, Rheinmetall-Borsig of Berlin, and Friedrich Krupp AG of Essen.

An armored vehicle of the 15-ton class was required. The situation of the fighting compartment in the front part of the vehicle moved the powerplant to the rear. This raised essential questions. The rear-engine design offered the following advantages: the rising heat from the motor (air vibration) did not interfere with targeting. The driver, seated far to the front, had a fairly good view. The powerplant was also well protected from shots. The only disadvantage, what with the drive sprockets being in front, was a long power train.

**Generaloberst Heinz Guderian, creator of the German Panzer troops.**

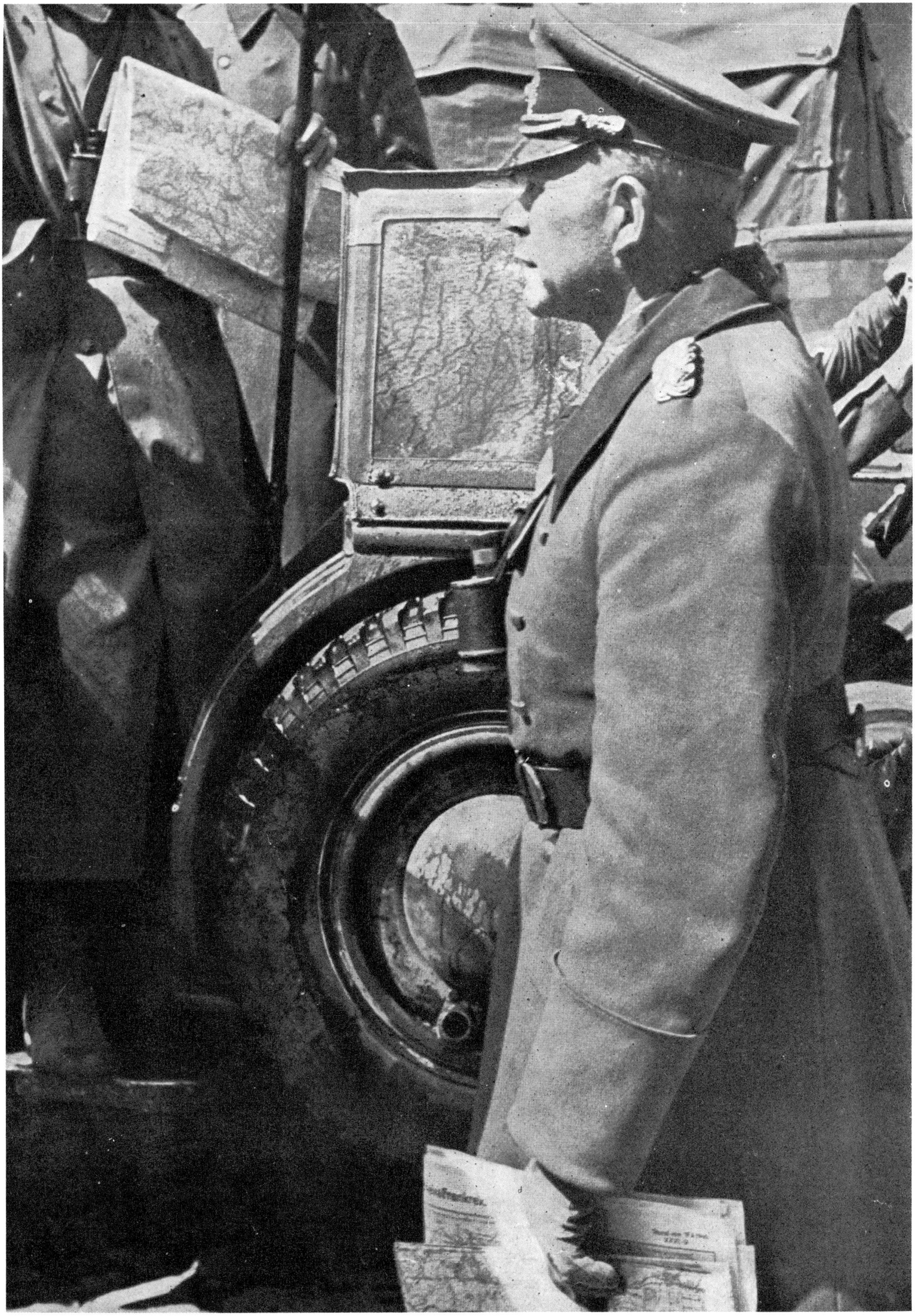

**A prototype built by the Friedrich Krupp firm in Essen. The vehicle was designated "MKA" was extended for export sales.**

The question of front or rear drive was examined critically. The following advantages of rear drive were cited: The tracks were only taut from the first road sprocket to the drive sprocket behind. Thus the tracks could flex in non-taut places and thus cause less wear to the joints of the tracks. The short power transmission from a rear-mounted powerplant was also an advantage. In addition, the drive sprockets were less exposed to enemy fire.

On the other hand, there were the following disadvantages: Long transmission lines were necessary for the steering system. In addition, the tracks, having picked up dirt, immediately reached the drive gears, which would cause greater wear.

The advantages of front drive were chiefly the self-cleaning of the tracks before they reached the drive sprockets, causing less wear to the tracks and drive gears. In addition, the steering lines could be kept as short as possible.

A disadvantage was the flexing of the tracks in a taut condition, which would result in greater wear.

In addition, this layout required a comparatively longer power transmission from the rear engine.

In terms of turning ability, it was stated that a short, wide tank was easier to control, as the motive power was applied to the large lever, while opposing ground pressure affected the short lever. In this layout too, there was greater security against tipping to the side.

In addition, better ground clearance, ditch-crossing, climbing and wading ability were constantly being

**The Daimler-Benz AG built the first pre-series vehicles of the "ZW" series.**

sought, and the constructors were expected to attain them.

The first prototypes were available at the end of 1935. On the basis of experience, and because of the urgency of this project, the firm of Daimler-Benz was contracted with for the series production of this vehicle, while Rheinmetall-Borsig in Düsseldorf was entrusted by the OKH with developing the rotating turret.

There were essential disagreements from the start as to the armament of the "Panzer III." While the Army Weapons Office and the Artillery Inspection considered a 37mm gun sufficient, the Motorized Troops Inspection immediately advocated a 50mm gun. Since the infantry had already been equipped with the 37mm antitank gun, and since only one armor-piercing weapon was supposed to be built for the purpose of uniformity, the construction of a more powerful weapon was rejected. Looking ahead, though, it was possible to give the turret's turning circle a large enough diameter to allow a bigger gun to be installed later.

**Panzerkampfwagen III (37mm), Ausf. A**

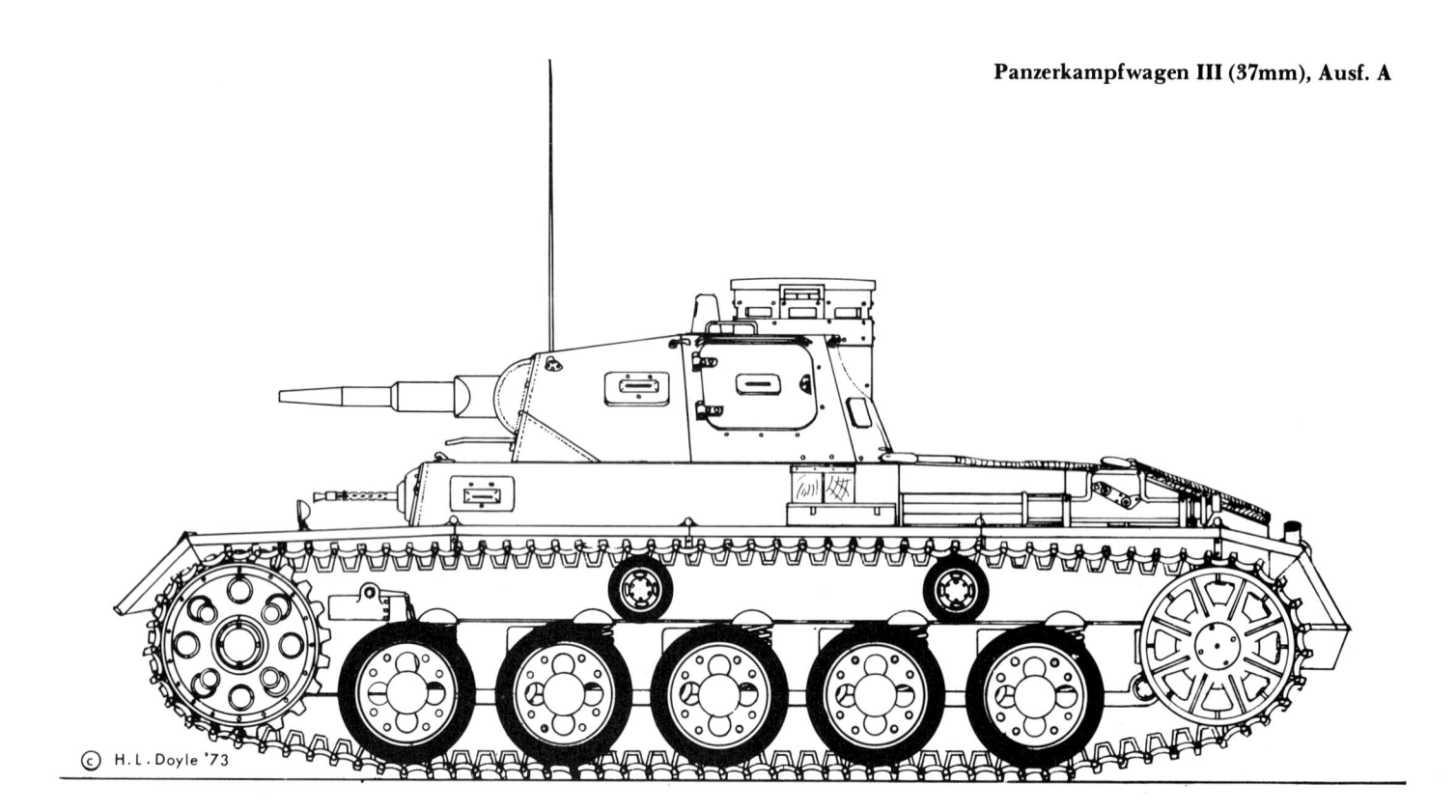

© H.L. Doyle '73

Originally, the turret designed by Rheinmetall-Borsig was fitted with a 37mm antitank gun, which allowed a muzzle velocity of 760 meters per second. Two MG 34 machine guns were installed in a special mantelet near the gun mantlet. A traverse of 360 degrees, an elevation value of -10 to +20 degrees was attained. Another MG 34 was installed in the bow of the vehicle.

Daimler-Benz equipped the first prototype vehicles with these rotating turrets. Ten of these "ZW" vehicles (chassis numbers 60 101 to 60 110), designated Panzerkampfwagen III (37mm) (Ausf. A), arrived for troop testing in 1936. Daimler-Benz designated the vehicles as Typ "1/ZW." Their gross weight amounted to 15.4 tons with 14.5 mm armor all around.

The tank itself consisted of the chassis, the armored superstructure attached to it, and the turret. The armored hull was designed to carry the chassis. It consisted of a closed front box, the open central hull, and the open rear compartment, formed of numerous pieces of armor, of different thickness, welded together. The sidewalls were strengthened by transverse members. Likewise the rear wall was reinforced to carry the axles of the idler wheels. The engine and drive train were mounted in the armored hull. The engine, a Maybach "HL 108 TR" 12-cylinder carburetor powerplant, was installed in the rear section of the hull, mounted three ways. At the sides of the engine were the radiators with ventilators and, on the left side of the vehicle, protected by a bulkhead separating it from the engine compartment, a fuel tank with 300-liter capacity.

A bulkhead with openings for the power transmission and electric lines separated the engine compartment from the fighting compartment.

A door in this bulkhead provided access to the engine compartment from the fighting compartment. The driveshaft led from the motor via the main clutch through a tunnel in the fighting compartment to the bow and the transmission (5-speed ZF "SFG 75"). To the front of the transmission gears was attached the bevel drive with the steering controls.

From the steering system, two shafts led left and right to

One of these vehicles in service with the 1st Panzer Division in Poland, 1939.

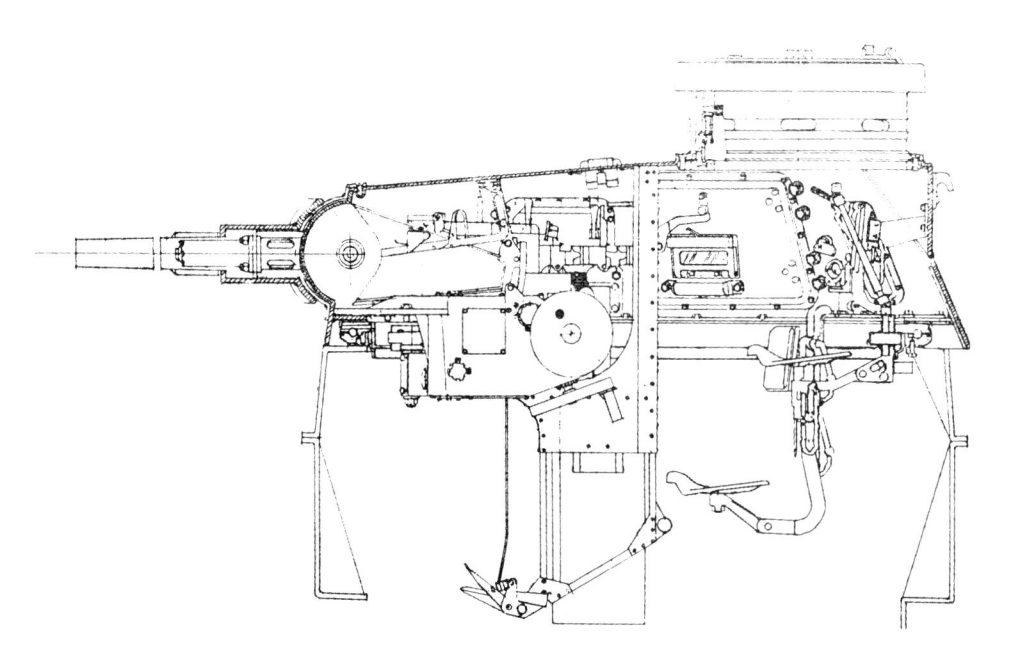

The turret of the A type was developed by the firm of Rheinmetall. This is a cutaway drawing showing the 37mm primary armament.

the two steering brakes and the two side rods. The steering brakes were inside, the side rods, which carried the drive sprockets, were attached to the outside of the hull sidewalls. Next to the transmission were the driver's seat on the left and the radio-operator's seat on the right. At the rear, the idler wheels were mounted on adjustable crank axles. Between the drive and idler wheel on each side were five road wheels, sprung by coil springs. Two return rollers were planned for each side, above the road wheels.

The tracks were kept taut over all the wheels, driven by the drive sprockets and covered by the aprons above them. The superstructure of the hull armor consisted of the bow and rear armor. This was screwed onto the hull, and either section could be removed independently of the other.

In the front wall of the bow armor, at the driver's eye level, was a driver's visor, and a round mantlet was located in front of the radio-operator. In addition, two visors were built into the sidewalls of the bow armor. In the right sidewall there was also an opening for the radio antenna.

The rear armor protected the engine compartment. Two hatches provided access to the engine. The air inlet louvers for the cooling system were located at the back of the rear armor.

The turret, which could be turned 360 degrees by hand, was mounted on ball bearings. The turret was welded together out of several sheets of armor. The side panels were curved and provided considerable width at the front, where they met the front wall of the turret. The gun mantlet was located inside and mounted in the front wall of the turret.

In the left and right sidewalls of the turret there were a hatch and a visor on each side. The hatches were covered by one-piece hatch covers, the visors by flaps. On the rear of the turret top was the commander's cupola.

The turret armor was stiffened around its lower rim by the welded-on ring, which held the turret on its bearings.

The supports for the folding seats of the commander and the gunner were screwed onto the ring of the turret.

The loader's folding seat was attached to the rear bulkhead of the fighting compartment.

At the end of 1936 the 1st Panzer Division received three of these tanks for training purposes.

Meanwhile a series of tests of the "ZW" vehicles had been initiated, and the vehicles had been fitted with completely new running gear. It was an eight-wheel type, in which pairs of road wheels were mounted together in each swinging mount. Each group of four road wheels was sprung by one leaf spring. There were also three return rollers on each side now. The drive mechanisms of these vehicles remained unchanged. In the superstructure, only the shape of the commander's cupola had been changed. The gross weight of these vehicles amounted to 15.9 tons. Twelve of these "Typ 2/ZW", also known as "Panzerkampfwagen III (37mm) Ausf. B", were built, with chassis numbers from 60 201 to 60 215.

The third experimental series, Ausf. C, also called "Typ 3a/ZW", was finished in 1937 and 1938, as was the Ausf. B series. The 8-wheel running gear again had every two road wheels mounted in a double swinging mount. But now the two foremost and two rearmost road wheel pairs were each sprung by one leaf spring, and the middle four wheels by one larger leaf spring. The small leaf springs lay horizontally (chassis no. 60 301 to 60 315). Otherwise the vehicle was just like the Ausf. B.

In the "3b/ZW" Ausf. D pre-series version of Panzerkampfwagen III, the main armor was upgraded from 14.5 mm to 30 mm. The gross weight now added up to 19.8 tons. In the running gear, almost unchanged from Ausf. C, the smaller leaf springs were now set at an angle. A six-speed gearbox of the ZF SSG 76 type was used (chassis no. 60 221 to 60 225). The upgrading of the armor required the installation of a Fahrersehklappe 30 (driver's visor 30) and Kugelblende 30 (ball mantelet 30). The commander's cupola was changed again.

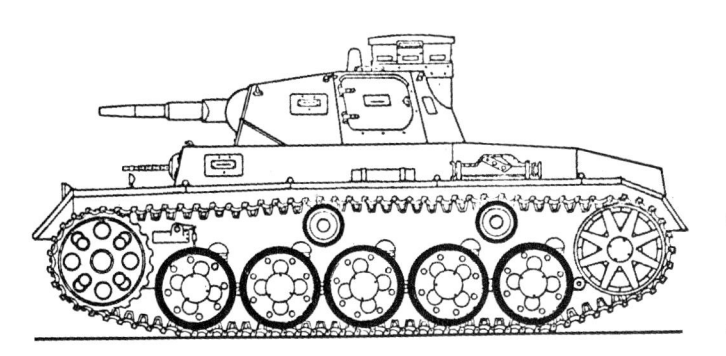

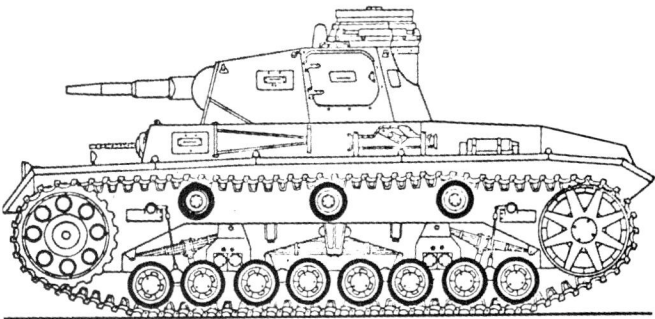

Ausf. A

Ausf. D

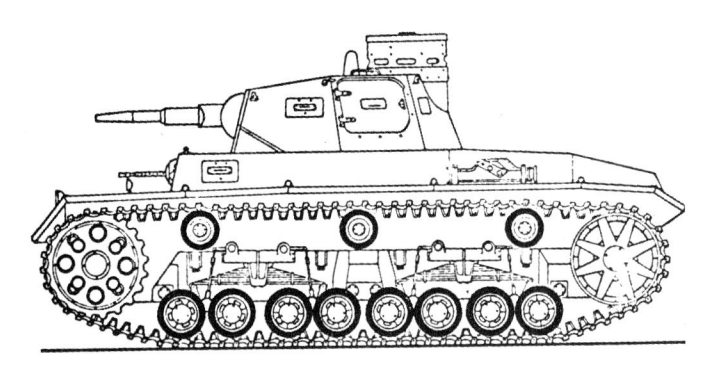

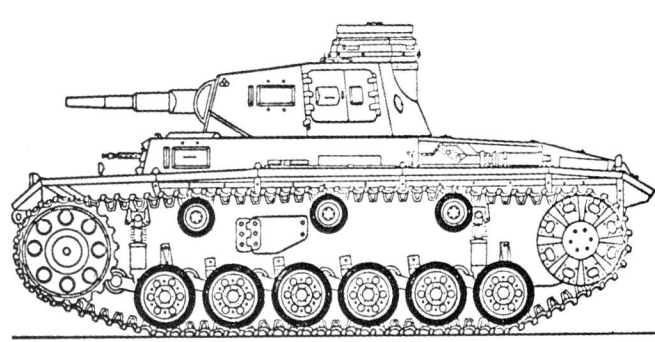

Ausf. B

Ausf. E

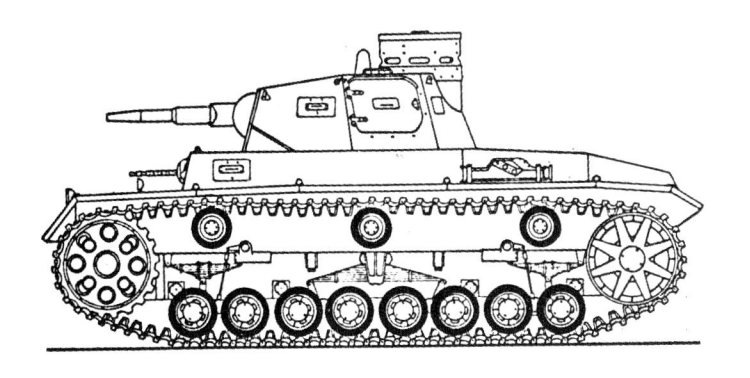

Ausf. C

The running-gear design changed considerably during the developmental period. These schematic drawings show the Ausf. A to E vehicles.

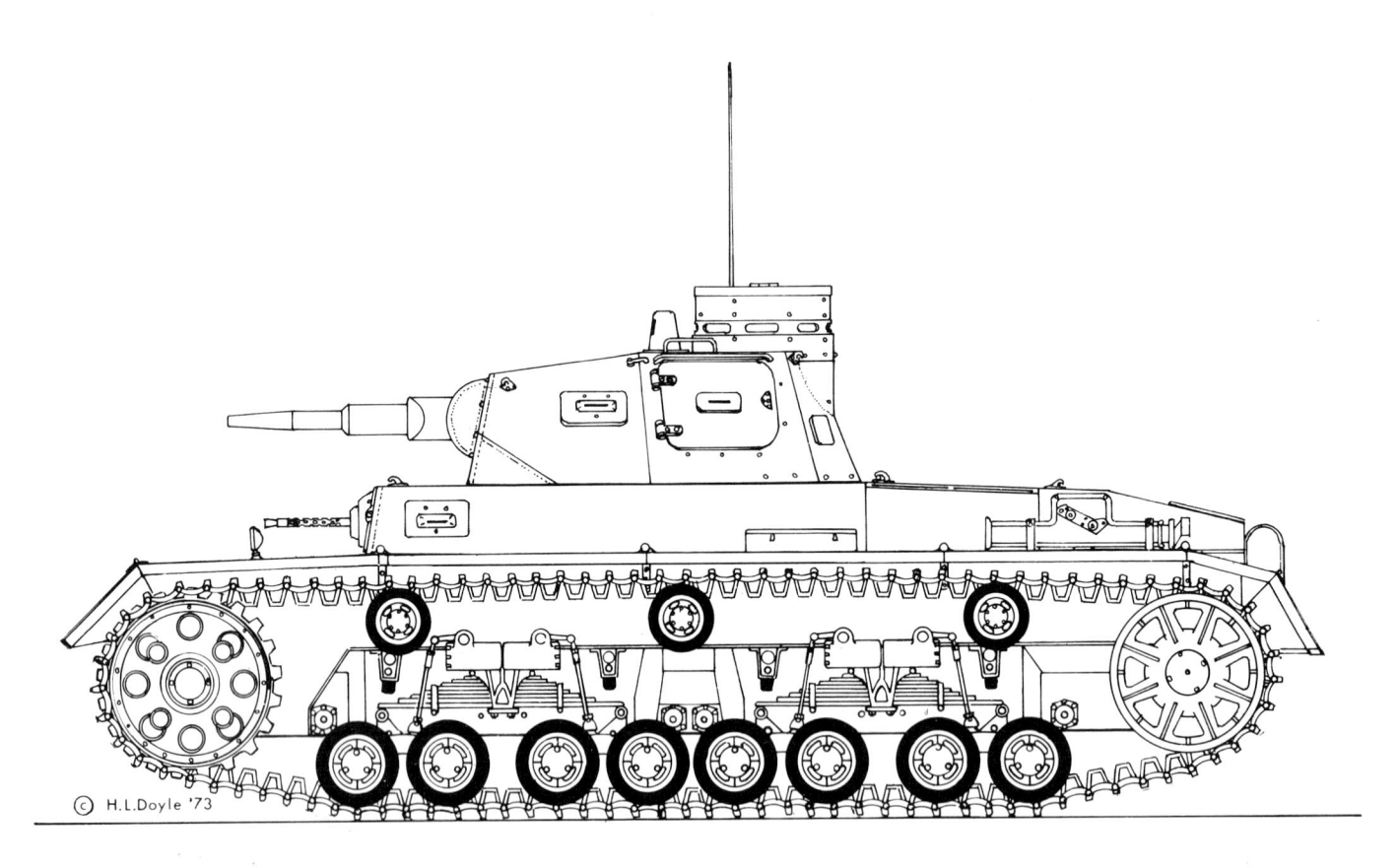

**Panzerkampfwagen III (37mm), Ausf B**

**A Ausf. B vehicle with the typical eight-wheel running gear. The technically costly design is easy to see.**

## The 1938 Series

All series of the "Panzer III" were equipped with the type series 2 and 3 directional gyrocompass or with the Model 8 directional gyro.

A variant of the Panzerkampfwagen III, Ausf. D, the Panzerbefehlswagen (armored command vehicle) developed by Daimler-Benz at Berlin-Marienfelde, appeared in 1938. These vehicles were intended for the commanders of the Panzer units. In their external form these vehicles much resembled the tanks. The crew likewise consisted of five men, these being the commander, the adjutant (simultaneously the machine-gunner), the tank driver and two radio-operators. The turret was screwed onto the superstructure armor and did not turn. The primary armament was actually a dummy. In addition to the turret machine gun, an MG 34 was mounted ahead of the driver in a type 30 ball mantelet. In addition, the crew had three machine pistols available. Two rod antennas (1.4 and 2 meters long), a "P" crank mast (9 meters long when extended), and a star antenna that went with it, along with a frame antenna attached to the stern

**Panzerkampfwagen III (37mm), Ausf. C**

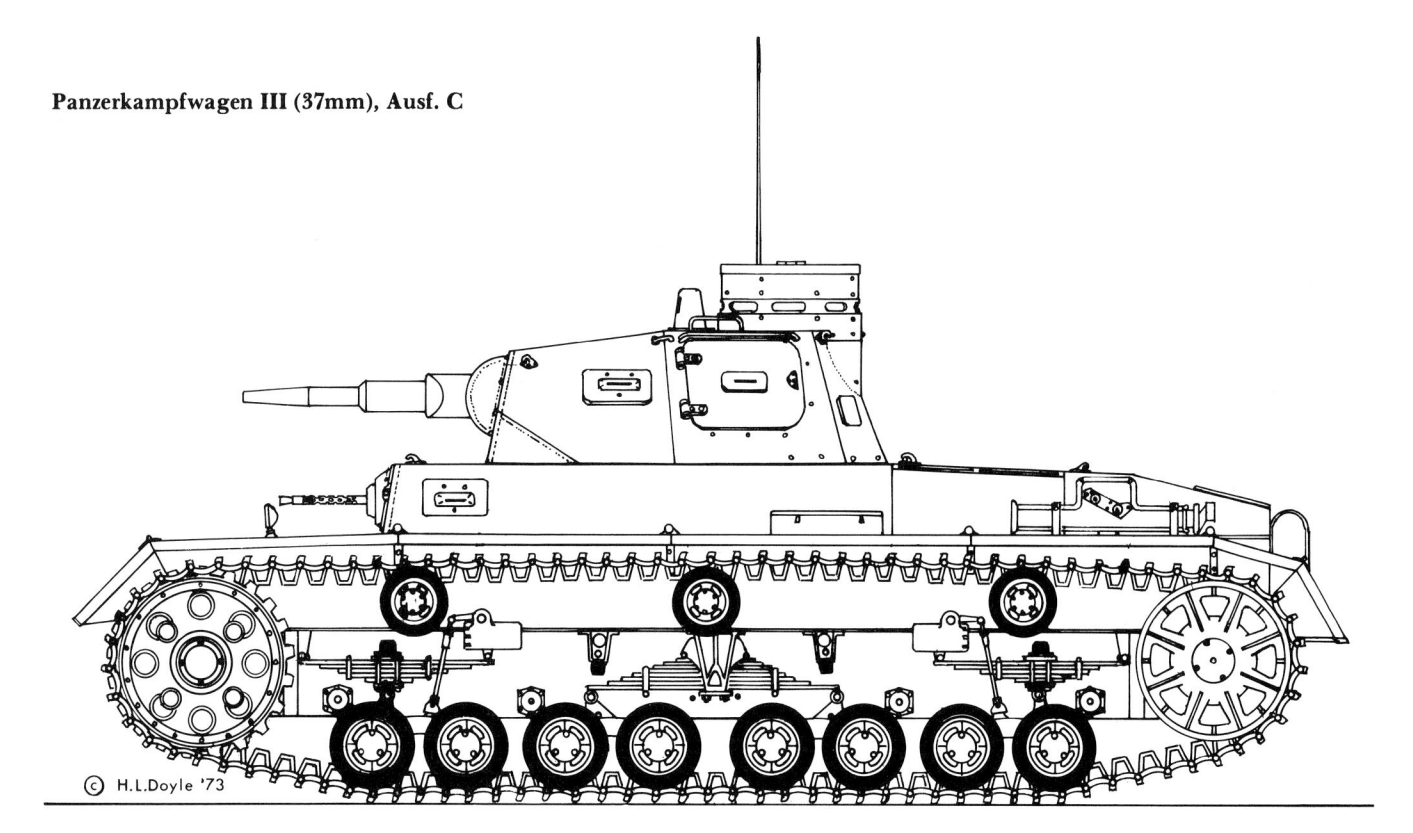

© H.L.Doyle '73

armor completed the radio equipment. Three versions of this vehicle were delivered. They were the Sd.Kfz.266, equipped with one Fu 6 and one Fu 2 radio, the Sd.Kfz.267 with one Fu 6 and one Fu 8, and the Sd.Kfz.268 with one Fu 6 and one Fu 7. Cable drums and guy-wires were planned for. The gross weight amounted to 20 tons. The official vehicle designation was Panzerbefehlswagen III, Ausf. D1, formerly Ausf. A. Daimler-Benz used the type designation "3c/ZW", and the chassis numbers were 60 341 to 60 370.

In 1939 there appeared the final production vehicle, the "Type 4/ZW", which was also called "ZW 38" by Daimler-Benz. A completely new running-gear design was used for this vehicle.

As before, both final drive systems, two bevel gears with a ratio of 4:1, were mounted on the outside of the hull, each in a housing made of armor plate. The drive sprocket was made of cast steel and screwed to the flange axle of the final drive. Two interchangeable sprockets, each with 21 teeth, meshed with the tracks.

The idler wheel consisted of a hub onto which two wheel discs were welded. These were equipped with hardened guiding rings to direct the teeth of the tracks laterally. The idler wheel turned on roller bearings around a crank axle which was mounted in the rear of the hull. The tension on the tracks was created by adjusting this crank axle.

The three return rollers (size 310 x 70-302) turned on fixed pivots mounted in special blocks. They took the form of double wheels equipped with removable rubber tires, between which the teeth of the tracks were directed laterally by guiding rings.

The road wheels, 6 on each side (size 520 x 95-398), were likewise formed as double wheels and consisted of two steel plates welded onto a hub, with rims carrying full rubber tires. The teeth of the tracks were also guided laterally here by inner guiding rings. Every road wheel turned on an axle that was pressed into a swinging arm mounted on the hull. A torsion bar spring — a bar made of spring steel, with two toothed heads — contacted the swinging arm with one head, while the other was held by a nut near the mount of the swinging arm on the other side, the opposite side of the hull. The outer ends of the swinging arms were limited in their upward movement by

The Ausf. D still had a slightly modified 8-wheel running gear. One of these vehicles is shown here during the 1939 Polish campaign. The shield of the folding radio antenna can be sen on the right track cover. The turret had a new commander's cupola, not used on the Type C.

This front view of the Ausf. D shows the driver standing in his entry hatch.

Panzerkampfwagen III (37mm), Ausf. D (Sd.Kfz.141).

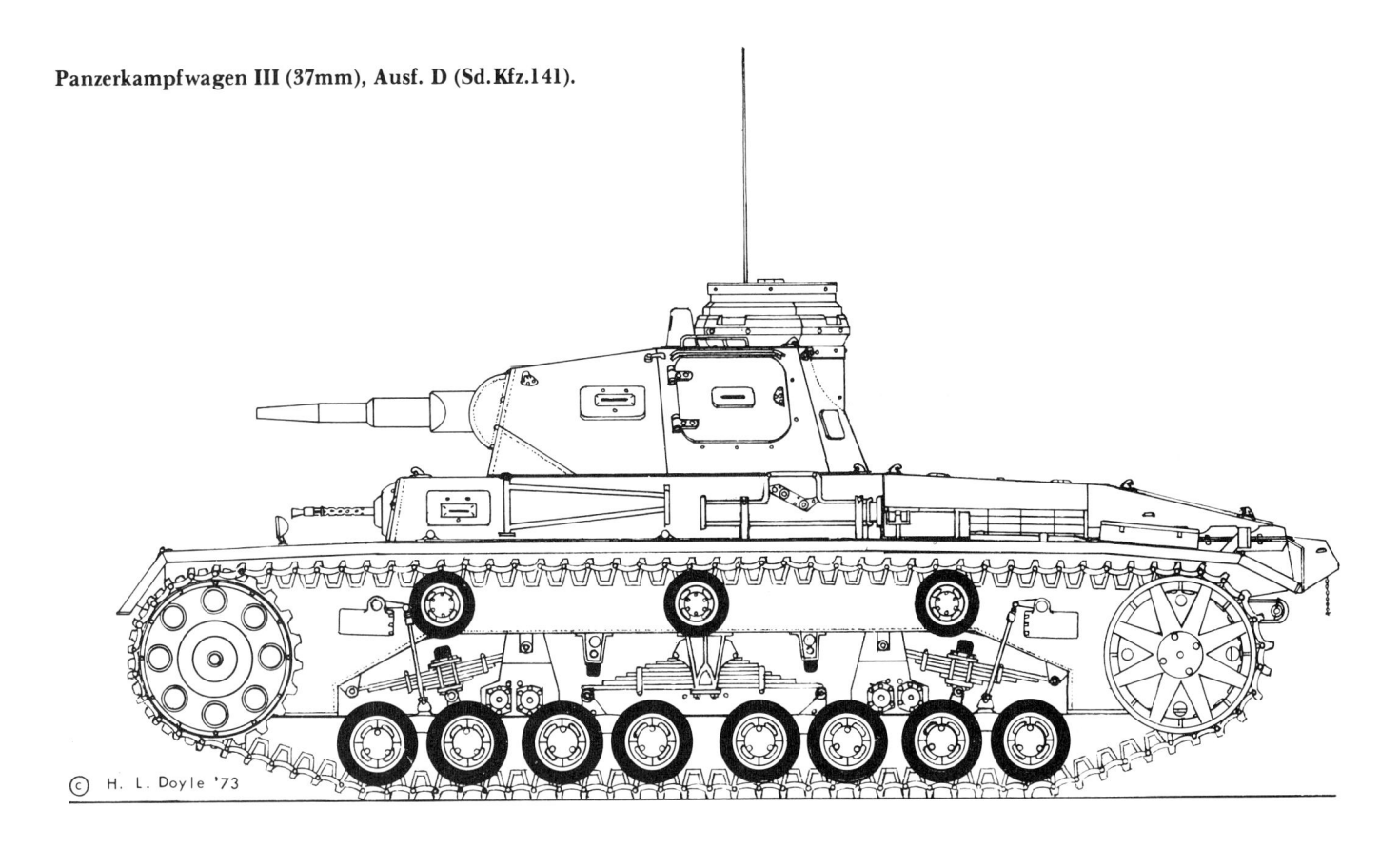

© H. L. Doyle '73

An eight-wheel chassis of Ausf. D, compared with the "ZW 38" made by Daimler-Benz. The simplification of the running gear after the adoption of torsion-bar suspension is obvious.

19

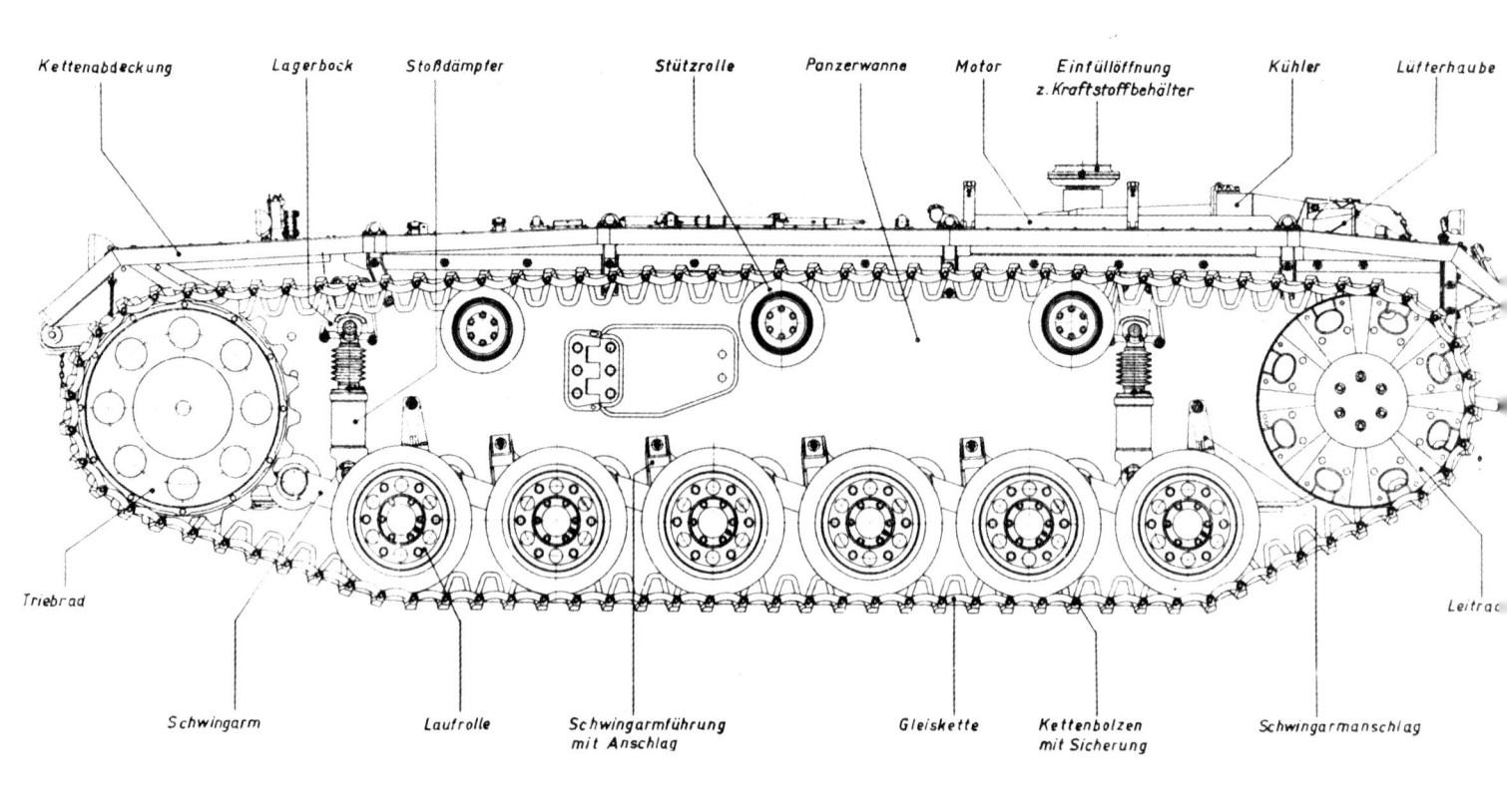

Kettenabdeckung | Lagerbock | Stoßdämpfer | Stützrolle | Panzerwanne | Motor | Einfüllöffnung z. Kraftstoffbehälter | Kühler | Lüfterhaube

Triebrad | Leitrad

Schwingarm | Laufrolle | Schwingarmführung mit Anschlag | Gleiskette | Kettenbolzen mit Sicherung | Schwingarmanschlag

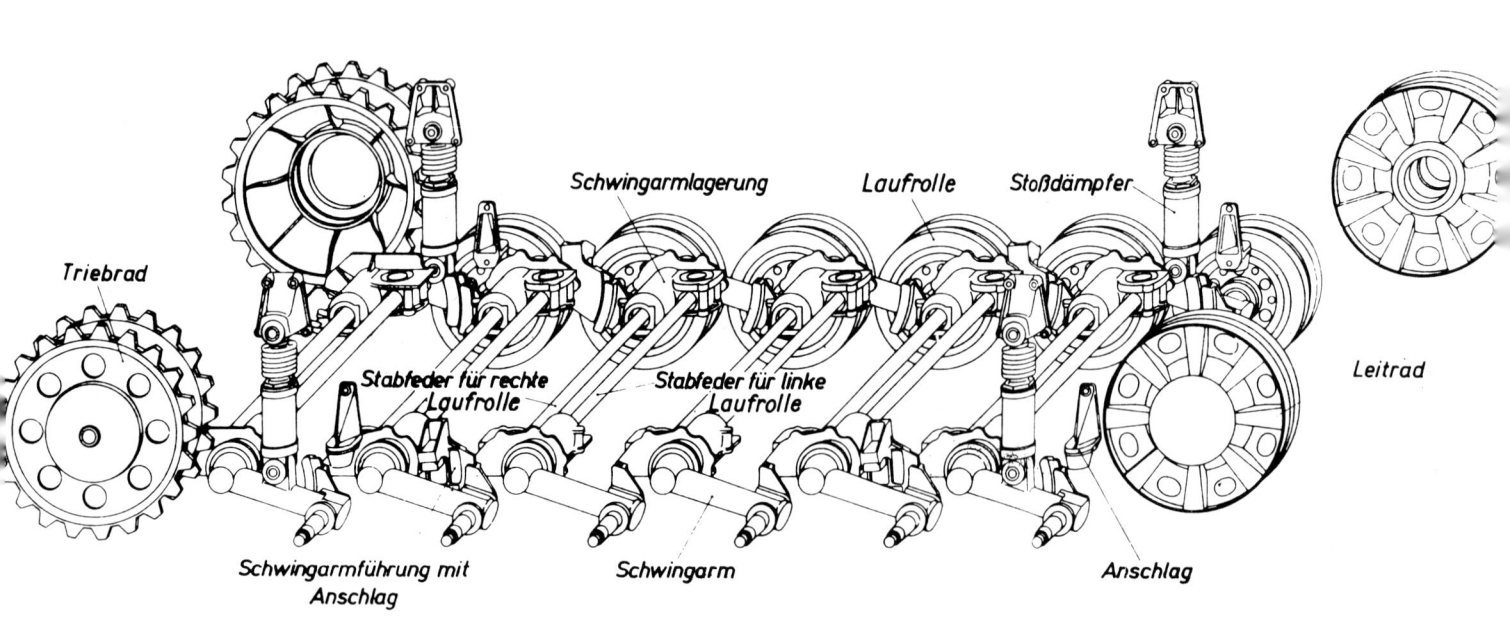

Triebrad | Schwingarmlagerung | Laufrolle | Stoßdämpfer | Leitrad

Stabfeder für rechte Laufrolle | Stabfeder für linke Laufrolle

Schwingarmführung mit Anschlag | Schwingarm | Anschlag

Section A-B

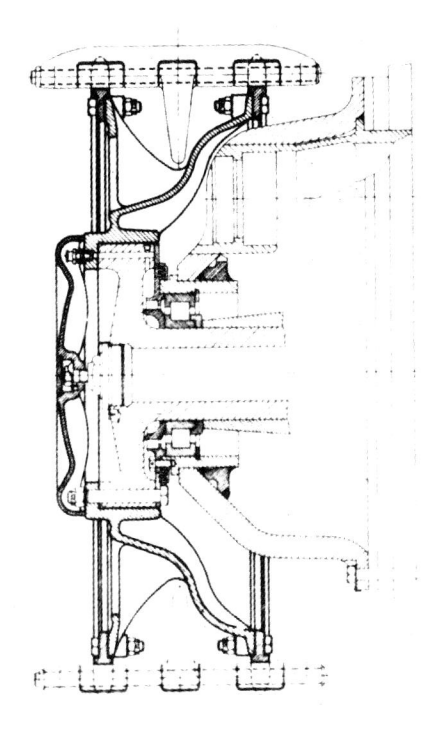

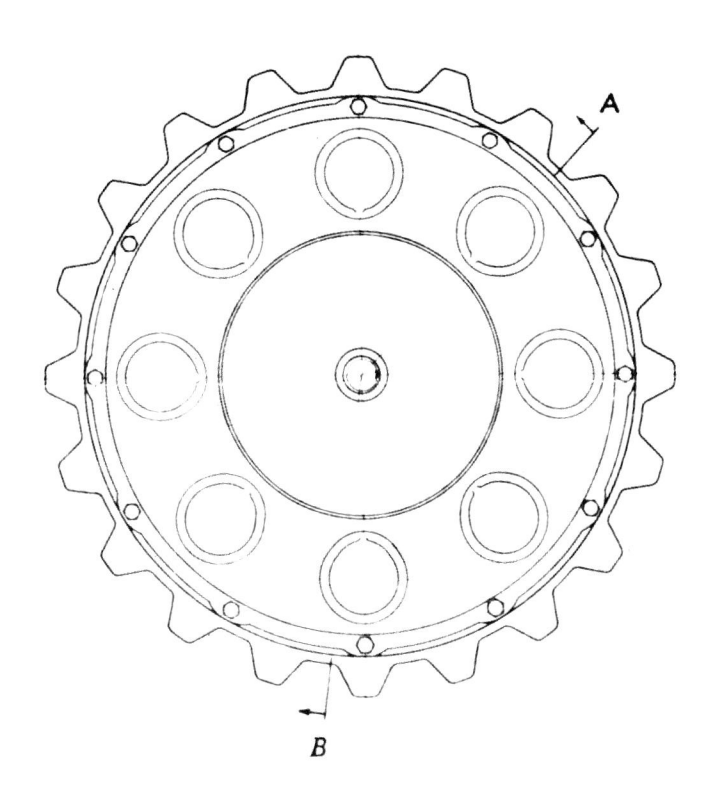

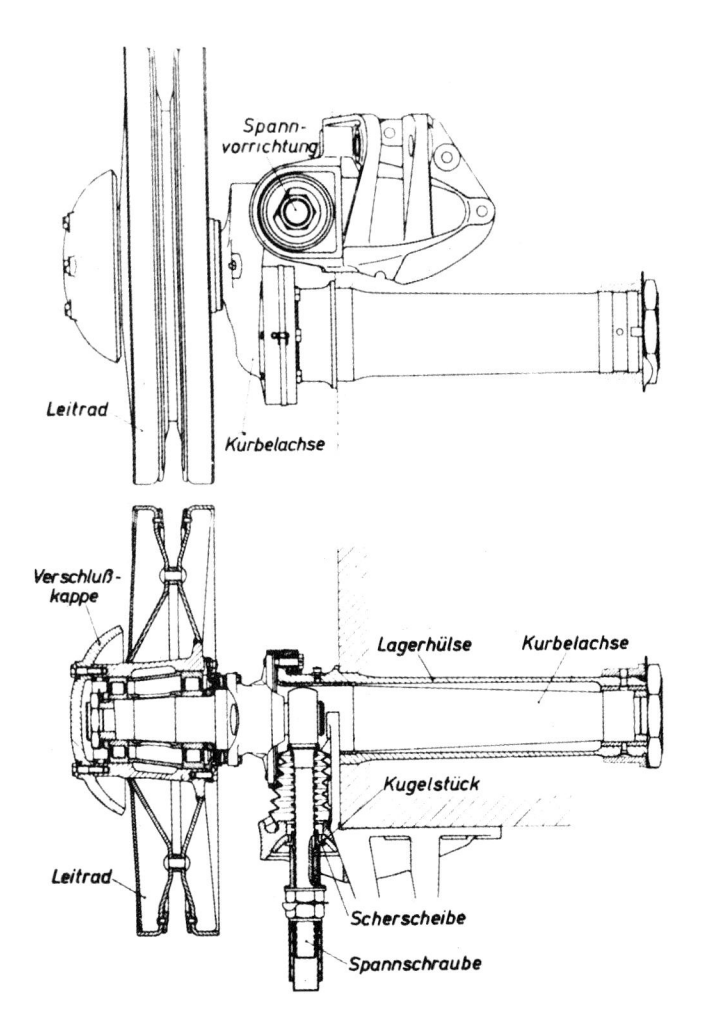

Pictures show details of the drive sprockets, as well as the idler wheels with the track-tensioning devices for the "ZW 38" chassis.

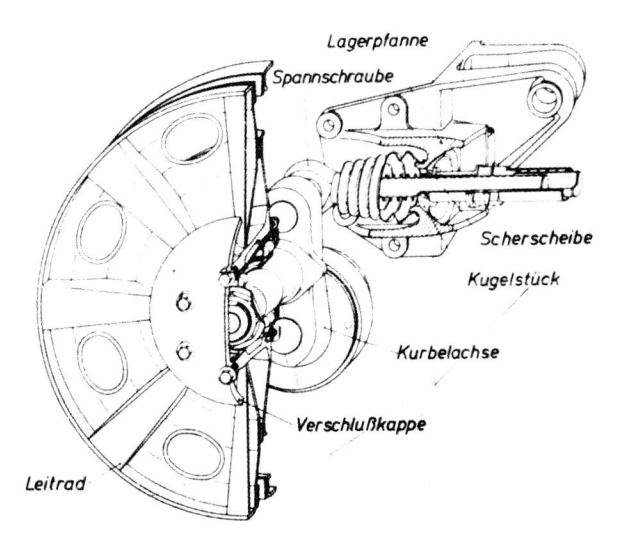

Spannvorrichtung
Leitrad
Kurbelachse

Verschlußkappe
Lagerhülse
Kurbelachse
Kugelstück
Leitrad
Scherscheibe
Spannschraube

Lagerpfanne
Spannschraube
Scherscheibe
Kugelstück
Kurbelachse
Verschlußkappe
Leitrad

21

rubber contact blocks. To absorb the lateral pressure, every swinging arm passed along a guiding rail attached to the armored hull.

To absorb pitching swing, the front and rear swinging arms on each side of the vehicle were each equipped with a hydraulic shock absorber working on one side.

The unlubricated tracks were composed of individual interlocking links held together by bolts. The track pitch amounted to 120 mm.

In technical terms, the larger Maybach Type HL 120 TR engine with two shaft-driven magnetos was used. From it a balance-weighted shaft carried the energy to the main clutch.

A Maybach Variorex SRG 32 8 145 pre-selector gearbox was used in the transmission. The individual gears were selected in advance, but the shifting itself was done manually by a low-pressure system, in which a valve was activated as soon as the clutch pedal was stepped on. The gearbox included ten forward speeds and one reverse gear.

The bevel gears and the steering gears were attached to the gearbox in a single housing. From the steering gears, the power passed to the left and right, via two shafts, to the steering brakes and the final drive.

The support and steering brakes were servo-operated inside drum brakes, each with two brake shoes. Power transmission for steering operated hydraulically.

In the superstructure, there were double entry hatches in the sidewalls of the turret. In some of the Ausf. E tanks there was no visor to the right of the radio-operator's seat. Production of this Panzerkampfwagen III, Ausf. E continued until February of 1939; the chassis numbers were 60 401 to 61 000 (without 60 501 through 60 545).

The army regulations published on September 27, 1939 included the information that the "Panzerkampfwagen III (37mm) (Sd.Kfz.141)" was being declared ready for introduction and use on the basis of successful troop testing.

The following firms took part in the large-series production of these vehicles:
Altmärkisches Kettenwerk GmbH (ALKETT), Spandau works, for assembly, Falkensee works for chassis construction, Daimler-Benz AG, Berlin-Marienfelde works,

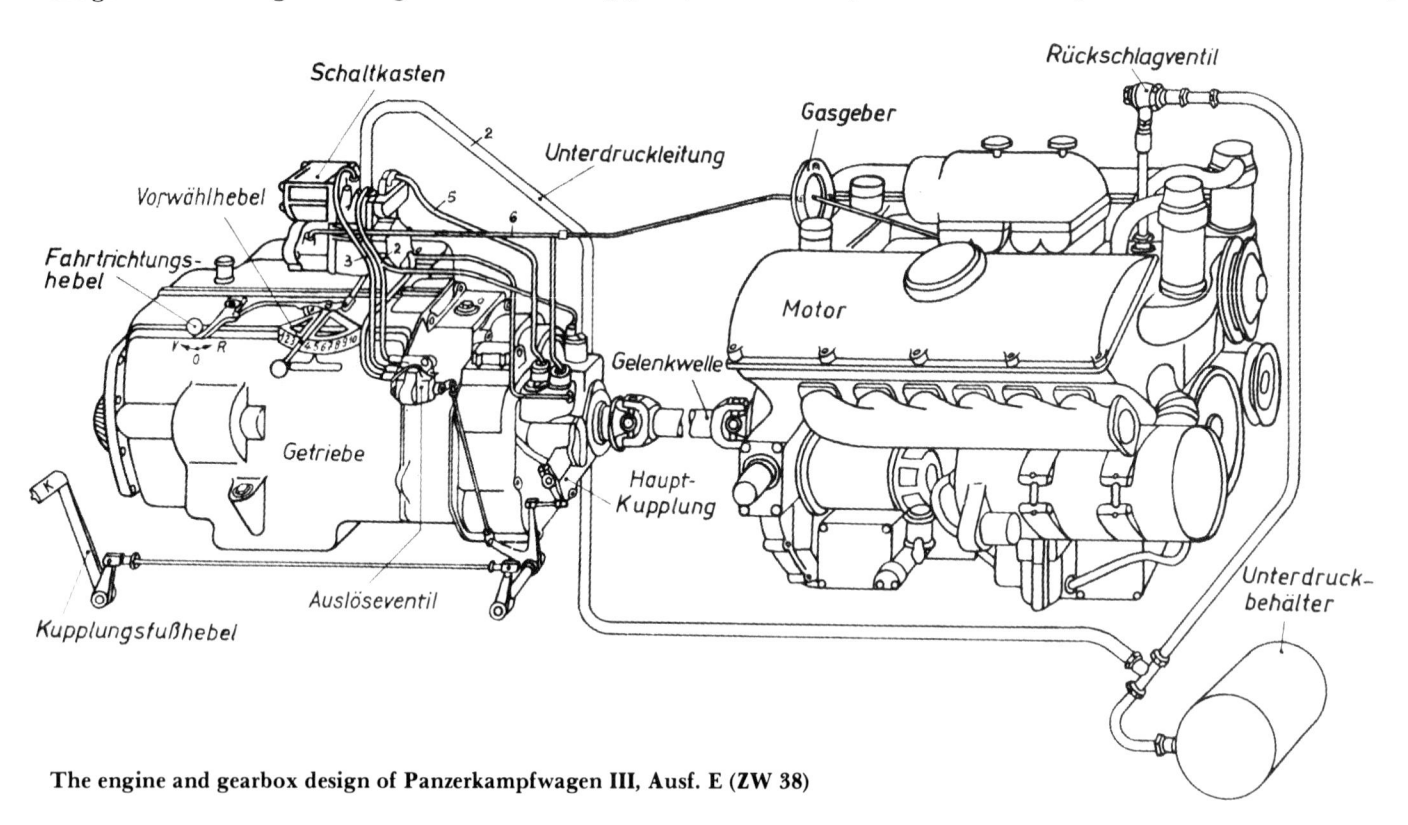

The engine and gearbox design of Panzerkampfwagen III, Ausf. E (ZW 38)

Side and front views of the "4/ZW", the Ausf. E Panzerkampfwagen III. This vehicle did not yet have a steering-brake vent.

**Panzerkampfwagen III (37mm), Ausf. E (Sd.Kfz.141)**

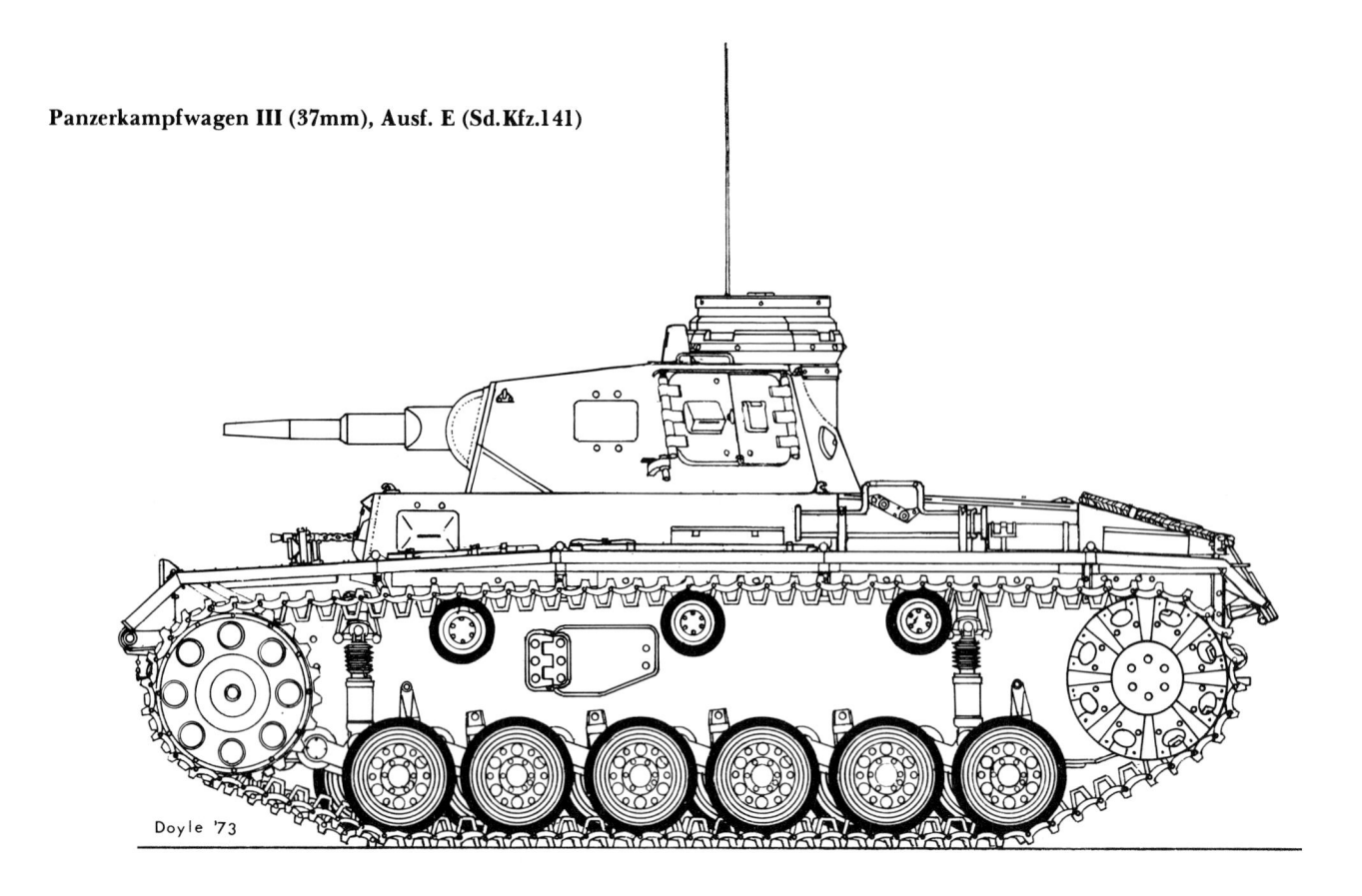

Doyle '73

Fahrzeug- und Motorenbau GmbH (FAMO), Breslau works, Henschel & Sohn AG, Mittelfeld-Kassel III works, Maschinenfabrik Augsburg-Nürnberg AG (MAN), Nürnberg works, Mühlenbau und Industrie AG (MIAG), Amme works, Braunschweig, Waggonfabrik Wegmann AG, Kassel works, Maschinenfabrik Niedersachsen-Hannover (MNH), Hannover-Linden works.

Of the firms listed here, the Alkett firm produced the greatest number of the "ZW" vehicles. The Wegmann Waggonfabrik AG of Kassel was involved from 1937 to mid-1942.

Henschel & Sohn produced the first five Panzer III tanks of its own manufacture in April of 1939. In October of 1942 a high production output of 63 of these vehicles was attained. Panzer III production by Henschel ended in November of 1942.

The Mühlenbau- und Industrie AG Amme works in Braunschweig (MIAG) produced Panzer III tanks from 1941 to 1943 in the following quantities:

| Month | 1941 | 1942 | 1943 |
|---|---|---|---|
| January | 6 | 40 | 80 |
| February | 20 | 42 | — |
| March | 20 | 45 | — |
| April | 25 | 50 | — |
| May | 25 | 50 | — |
| June | 25 | 55 | — |
| July | 25 | 60 | — |
| August | 30 | 70 | — |
| September | 32 | 70 | — |
| October | 31 | 70 | — |
| November | 35 | 70 | — |
| December | 35 | 80 | |

The total, then, was 1091 Panzer III vehicles. The Maschinenfabrik Augsburg-Nürnberg produced a total of 794 "ZW" vehicles from 1939 to 1943. The Maschinenfabrik Niedersachsen-Hannover delivered its quota of 30 tanks per month starting in 1940. FAMO in Breslau and Daimler-Benz in Berlin-Marienfelde, according to information provided on November 1, 1940, produced an average of 26 "ZW" type vehicles per month.

As was the case with the Panzer I and II, the Deutsche Edelstahlwerke AG firm in Hannover-Linden played a major role in the production of armored hulls, superstructures and turrets. From 1936 to 1942 it delivered, among other things, the following parts to the assembly firms:

**One of these vehicles in service in France, 1940. At this time greater numbers of them reached the troops.**

A Panzer III stuck in the mud shows details of its antennas as well as the opened commander's cupola.

| Year | Hulls | Super-structures | Turrets |
|------|-------|------------------|---------|
| 1936 | 11    | —                | 5       |
| 1937 | 40    | 40               | 40      |
| 1938 | 106   | 104              | 110     |
| 1939 | 221   | 138              | 155     |
| 1940 | 415   | 438              | 450     |
| 1941 | 605   | 593              | 593     |
| 1942 | 556   | 554              | 554     |

From 1939 on, approximately 100 of the "4/ZW" vehicle were built; they were used as the primary equipment of the Panzer regiments. Again and again, the low production capability of the German tank industry made itself felt at this time, being able to turn out only small numbers of vehicles. The tanks underwent their baptism of fire in Poland and proved themselves well. On May 10, 1940 there were 349 tanks available to the German Panzer regiments for the attack on France.

A variant based on the Ausf. E of the Panzer III was the Panzerbefehlswagen III Ausf. E (Command Car), formerly Ausf. B (chassis numbers 60 501 to 60 545). This vehicle type, "4a/ZW", was equipped and utilized like the already mentioned Panzerbefehlswagen III, Ausf. D1.

By January 4, 1938 the Waffenamt (Weapons Office) had already been empowered to begin work on a further development of the Panzer III with a 5-centimeter gun. As before, the Daimler-Benz AG was responsible for the chassis and superstructure, while turret manufacture was taken over by the Friedrich Krupp AG. Plans were made to install a 50mm KwK L/42 gun with a muzzle velocity of 450 to 685 meters per second. The barrel length was 2103 mm, the maximum shot range 6500 meters. The gun had a traverse field of 360 degrees and an elevation range from -10 to +20 degrees. The weight of the weapon was 400 kp. In spite of these plans, the Ausf. F version of the Panzer III (designated "5/ZW" by Daimler-Benz), first produced in 1939, with chassis numbers from 61 001 to 65 000, was still armed with the 37mm tank gun.

The same was true of the immediately following Ausf. G, though the 37mm gun was installed in only some of these tanks.

Production of the "ZW" vehicles with the 37mm KwK ended with 12 vehicles in August 1940. Production of the "ZF" vehicles with the 50mm KwK began with 10 vehicles in June 1940.

The front wall of the bow armor in the Ausf. E to H was later strengthened by welding on an additional 30 mm armor plate. Ausf. E and F, plus some of the Ausf. G, were later reshaped and rearmed with the 50mm KwK L/42 and an MG 34. These tanks in which the 37mm guns were replaced by the 50mm type also used the "TZF 5a, vorl. 50mm" sight. Beginning with the 5th series, two cast steering-brake cooling vents were installed in the upper bow plate. In addition, the Maybach "HL 120 TRM" engine with "snapper" magneto ignition was used. The cover of the left-side signal flap on the turret roof was eliminated again.

Ausf. E and F had a commander's cupola screwed onto the rear part of the turret. It provided the tank commander with an entrance as well as a view. It consisted of a cylindrical mantle with five upper and five lower levers, a numeral circle and a two-piece hatch cover. The cylindrical mantle had five openings which could be closed completely or partly by moving the pairs of upper and lower levers.

The already mentioned Ausf. G of the "Panzer III" bore the Daimler-Benz designation "6/ZW" and chassis numbers 65 001 to 66 000. A new commander's cupola was installed in this model. It consisted of a cylindrical mantle, the five upper and lower levers, and five side covers, the number circle and the two-piece hatch cover. The cylindrical mantle had five openings that could be partly or completely closed by two of the adjusting levers. The sidewalls of the turret were lengthened to the rear, and the commander's cupola no longer cut into the rear wall of the turret, which simplified production significantly.

On November 1, 1940 the production schedule for the Panzerkampfwagen III was set at 108 units per month. As a result of production problems for the seventh series (which first appeared in October of 1940), though, only 96 units were finished. In all, 895 of the "ZW" vehicles were built in 1940.

The Ausf. H Panzerkampfwagen III, type designation "7/ZW", with chassis numbers 66 001 to 68 000, featured several significant improvements to the chassis. Although the 30 mm armor plate was still used, basic changes were made to the drive sprockets and idler wheels. The drive sprockets were made of cast steel and screwed onto the flange shafts of the final drive sprockets. They were made in spoked form. Two interchangeable gears with 21 teeth each meshed with the tracks. The new idler wheels were also spoked and had two rim rings welded on. The width of the tracks was increased from 380 to 400 mm (track type Kgs 6100/120). To decrease pressure on the drive sprockets, the front return rollers were located closer to the drive sprockets. The transmission of power from the steering levers to the support or steering brakes was done hydraulically in Ausf. H. The complex Maybach VARIOREX gearbox was also given up and replaced by ZF Type "SSG 77" transmission that featured six-speed drive with synchronization. The gears of the six forward speeds were obliquely toothed and took a firm hold. The reverse gear of the Ausf. H was secured by a knob with a wire cable, and later by a manually operated lever.

While the Fahrersehklappe 30 (Driver's Visor 30) in the Ausf. E and F was operated by two overlapping levers, the Driver's Visor 30 of Ausf. G and H consisted of one housing screwed onto the front wall of the bow armor and spanned by a push lever. The driver's optics consisted of two double-angled periscopes, type "KFF 1" in Ausf. E

As of Ausf. F, a steering-brake ventilator was present. These cast pieces were screwed onto the bow armor.

The rear of the "5/ZW" vehicle shows the arrangement of the exhaust system as well as the original attachment of the smoke-laying apparatus, which was operated from inside the vehicle.

This Wegmann factory photo shows the Panzerkampfwagen III, Ausf. 5/ZW, from above. The signal hatch near the gunner's position was now made without a hood.

The crew cleans the 37mm tank gun. The internal mantelet is easy to see.

While the Ausf. E and F were still fitted with the 37mm gun in the factory, the Ausf. G was fitted during production with the 50mm KwK L/42 gun. This picture shows vehicles made by Daimler-Benz; the third tank already has a 50mm KwK in an outside mantelet.

**Panzerkampfwagen III Ausf. G (Sd.Kfz.141)**

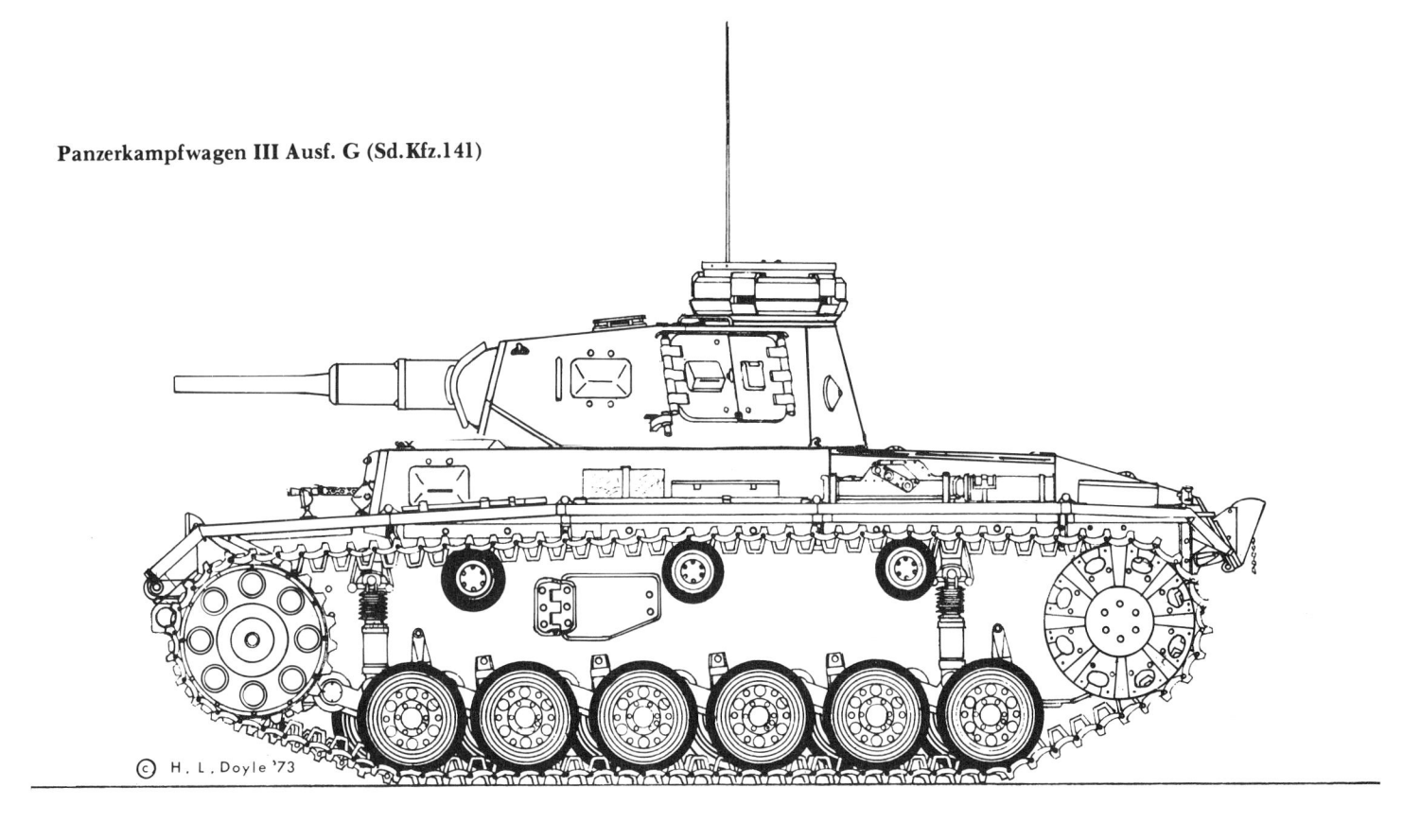

A Panzer III Ausf. F on the move. Note the folded-up front track covers, which made track self-cleaning easier.

This picture shows a Panzer III Ausf. E or F which was later rearmed with the 50mm KwK L/42. The front armor of the Ausf. E to H was later strengthened by screwed-on or welded-on 30 mm armor plate.

The "ZW" vehicles were now in large-scale production at several factories. This picture shows assembly at the Altmärkisches Kettenwerk GmbH in Berlin-Spandau.

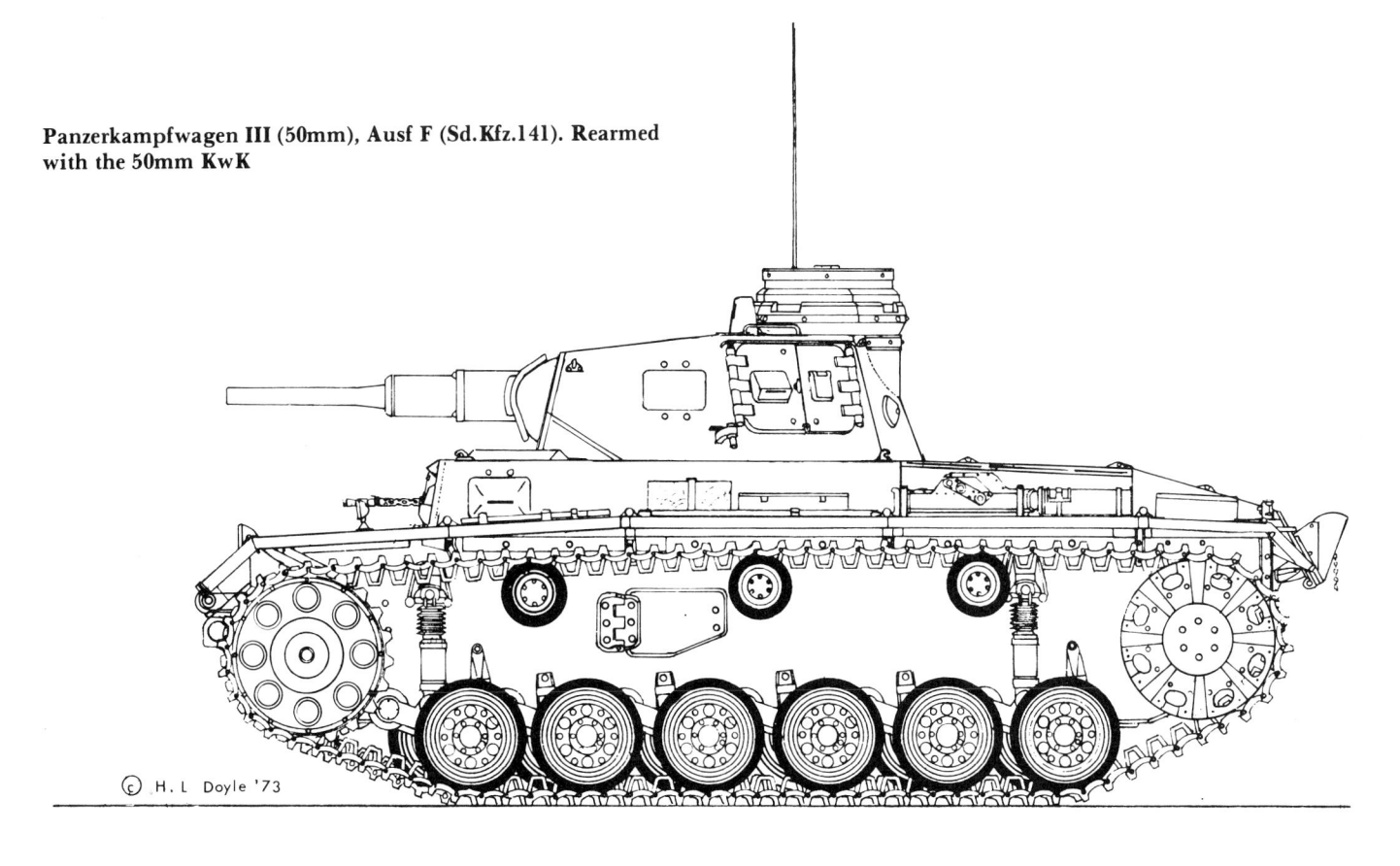

**Panzerkampfwagen III (50mm), Ausf F (Sd.Kfz.141). Rearmed with the 50mm KwK**

The Ausf. Type G was fitted with a new commander's cupola that no longer cut into the rear wall of the turret, as the turret sidewalls had been lengthened.

A Panzerkampfwagen III, Ausf. F with 50mm KwK in action in North Africa. There as in Russia, they played the leading role in tank battles.

The Type H (7/ZW) Panzerkampfwagen III had a modified chassis with new drive and leading wheels. The front return roller was also moved nearer to the drive wheel. Wider tracks were used. The basic armor was 30 mm thick, as before.

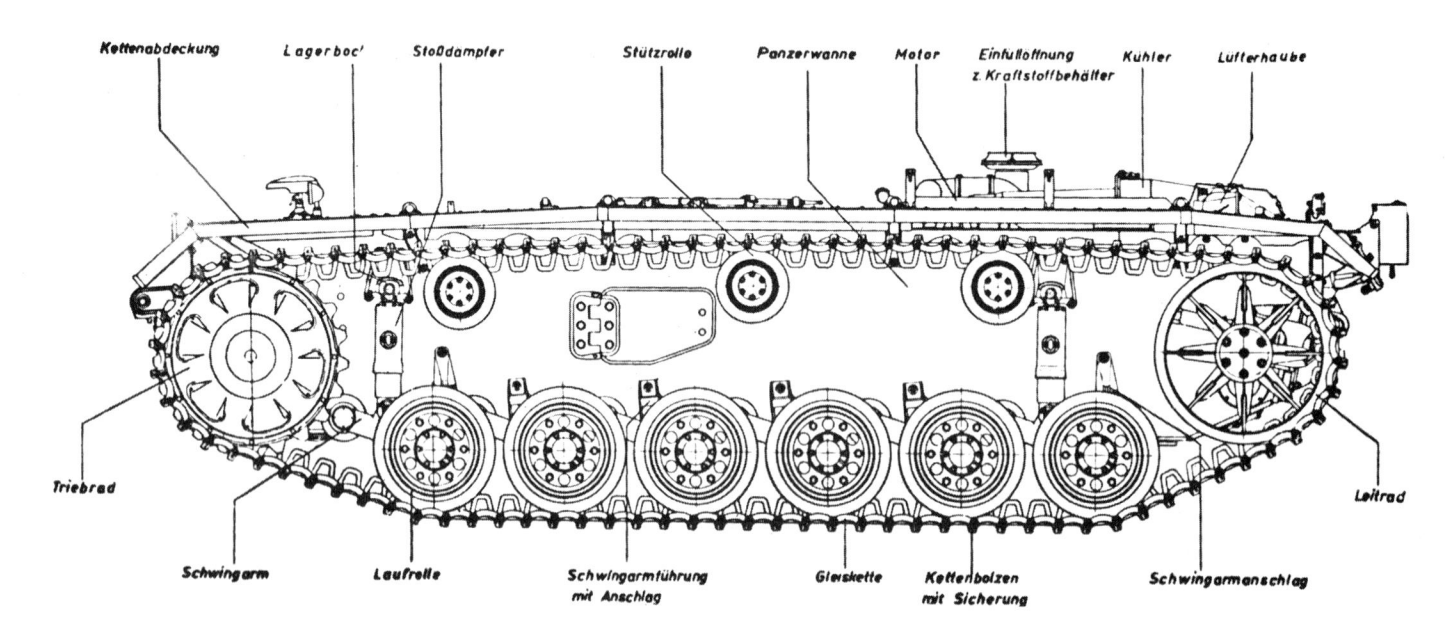

**Panzerkampfwagen III (50mm), Ausf. H (Sd.Kfz.141).**

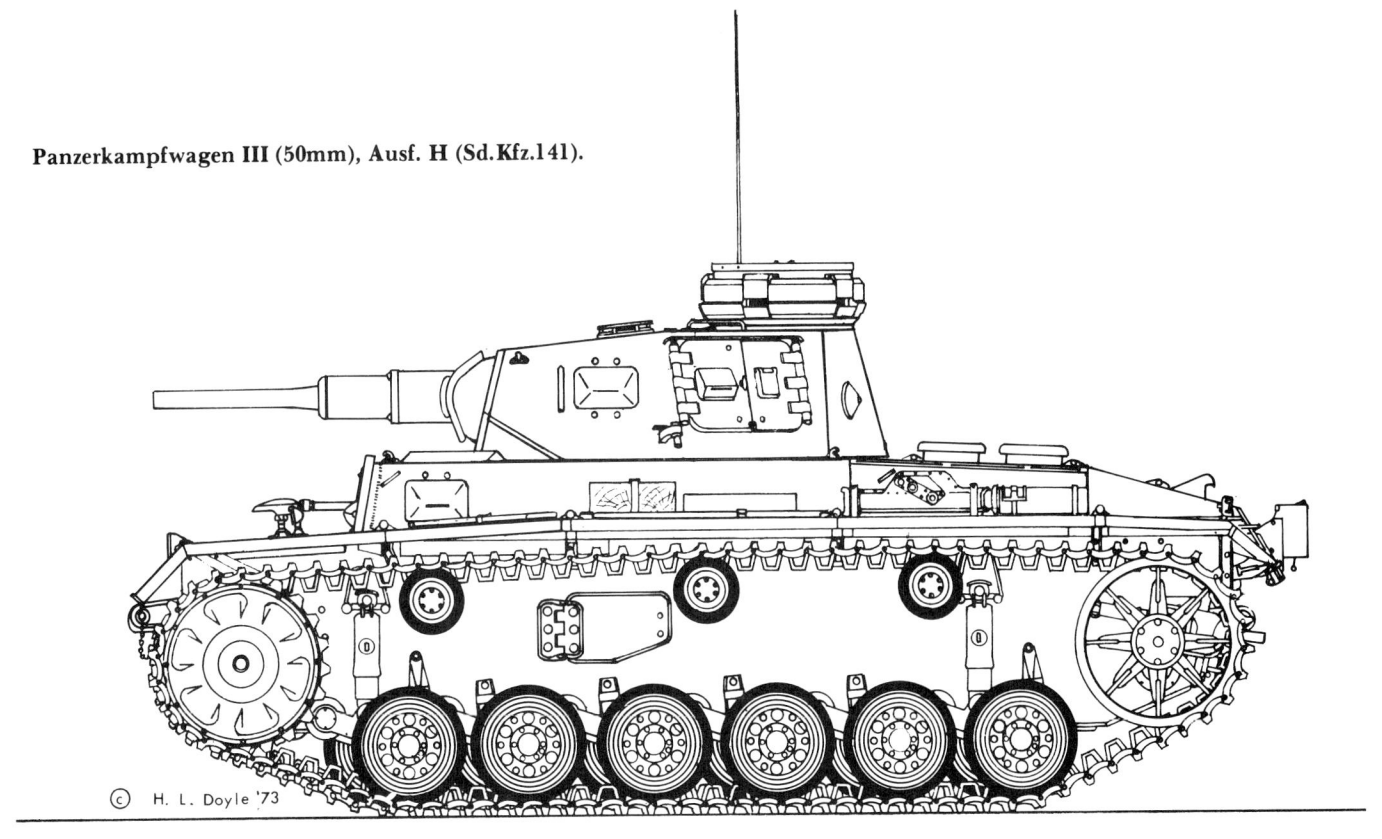

Schnitt A–B

Leitrad

Anhängegabel

Kurbelachse  Lagerhülse  Kurbelachse

The drawings show the detail changes to the Ausf. H drive sprocket and idler wheels.

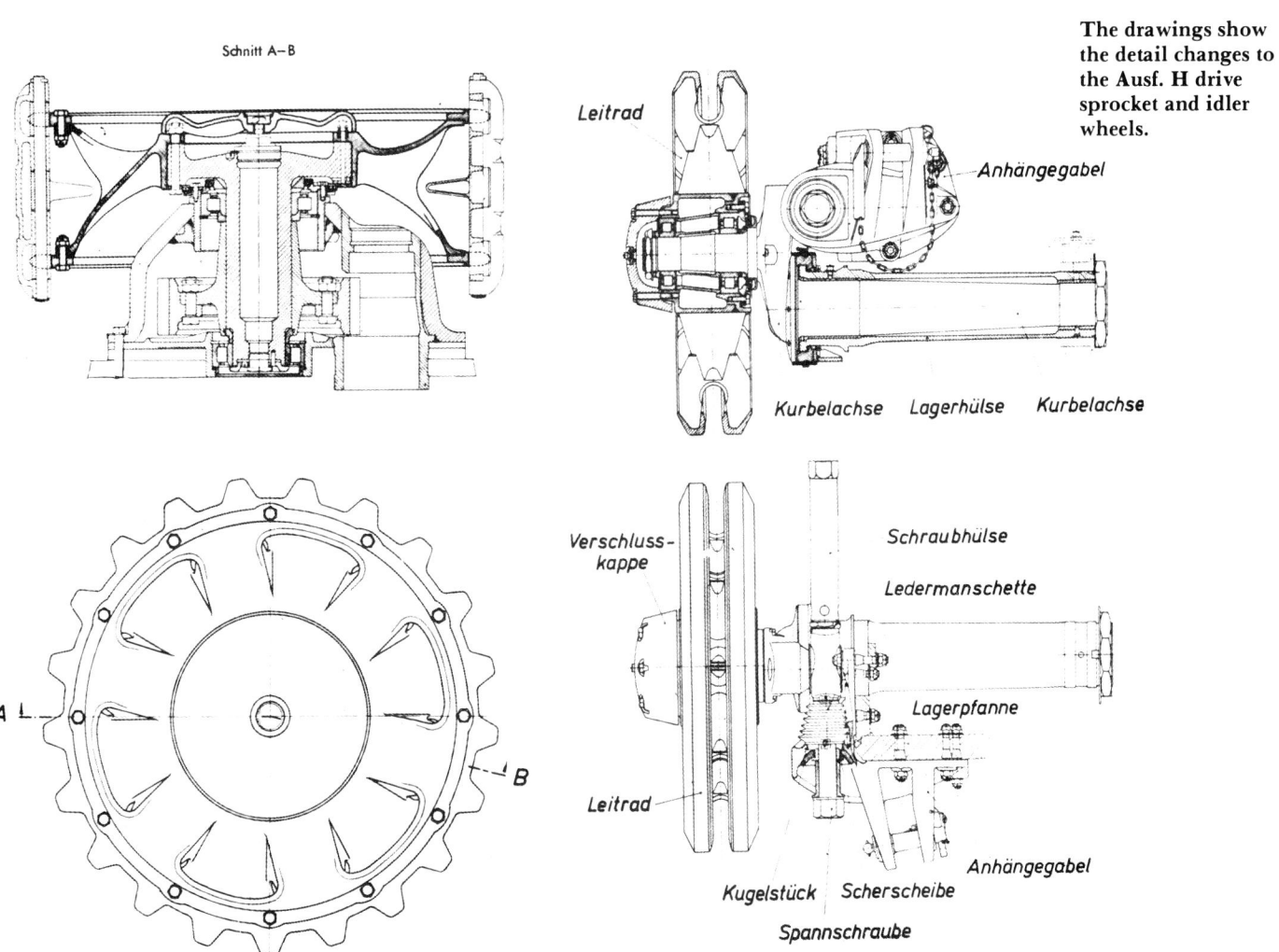

Verschlusskappe

Schraubhülse

Ledermanschette

Lagerpfanne

Leitrad

Anhängegabel

Kugelstück  Scherscheibe

Spannschraube

35

and F and "KFF 2" in Ausf. G and H. They afforded the driver an indirect view when the driver's visor was closed. Two viewing ports were drilled through the front wall of the bow armor above the driver's visor to provide a view.

The "Ball Mount 30" used in Ausf. E through H was located ahead of the radio-operator's seat on the right front wall of the bow armor. It consisted of a mantelet ball and its cover. The mantelet ball, mounted so it could be turned in any direction, had attachments on its back for an MG 34 machine gun with belt feed, as well as an aiming periscope (KFZ 2).

Some 300 vehicles of the Ausf. H had been made by 1941.

Hitler's order that all tanks brought back to Germany for overhauling should be updated to the latest technical state as far as possible resulted in many variations on the Ausf. H. There were vehicles with the old-type drive sprockets that had spacer rings between the geared wheels to allow the use of the wider tracks. A similar measure was taken to adapt the idler wheels. In all cases, the heavier weight of the tracks necessitated moving the front return rollers closer to the drive sprockets.

Corresponding to the Panzerkampfwagen III, Ausf. H was the Panzerbefehlswagen III, Ausf. H, formerly designated Ausf. C. The Daimler-Benz designation for this command vehicle was "7a/ZW", and its chassis numbers were 70 001 to 70 145. Production of the second series of these armored command vehicles (Ausf. E) had ended in March of 1940, and the first ten vehicles of the third Ausf. "H" version were delivered in November of the same year.

**This picture shows Ausf. G and H vehicles side by side modified for diving. The five-man crew stands before the tank.**

With the introduction of Ausf. H, overhauling of old vehicles resulted in hybrid running gear composed of old and new components. This picture shows an Ausf. E-G with new drive sprockets and modified return-roller attachment. The idler wheels are still the old type.

The German tanks that saw service in North Africa were given special tropical equipment intended to afford better cooling through the use of more effective air filters and relocated vents. Sometimes "felt-bellows" filters were used, mounted on the outside of the engine compartment and protected by armor plate. Even so, the usual piston lifetime in Africa amounted to only 2000 to 3000 kilometers. The vehicles fitted with this system bore the added designation of "Tp" (for "Tropical").

For Operation "Sealion", the proposed invasion of England, the Panzer Unit A, made up of volunteers from Panzer Regiment 2, was organized at Putlos in September and October of 1940. It was followed by two other special units, Panzer Battalions B and C. These three units later formed Panzer Regiment 18 of the 18th Panzer Division.

For Operation "Sealion", the invasion of England, Panzer III and IV were prepared for underwater action. This picture shows a Panzer III Ausf. F with closed openings and an air hose for the engine and the fighting compartment.

A Panzer III Ausf. F just before a diving test. The rods on either side of the body are marked to show the water depth for these tests. In the background is a loading ship, over whose forward ramp the vehicles went into the water.

The situation shortly before the tank dives. The buoy at the end of the air-supply hose is already in the water.

After "Operation Sealion" was given up, some of the diving tanks were turned over to the 18th Panzer Division. At the beginning of the Russian campaign in 1941, they crossed the Bug underwater. Instead of an air-supply hose, the vehicles used a steel pipe 3.5 meters long.

The "Panzer III and IV" tanks issued to these units were provided with wading capability by making the following changes: All openings, visors, flaps and the like were made watertight with rubber rims and cable tar. The turret entry hatches were fitted with inside bolts, and the air intakes for the engine were made to close fully. Rubber coatings were applied to the gun mantelet, the commander's cupola and the radio-operator's machine gun. In these coatings were explosive fuses that blew off the coatings after the vehicle emerged from the water and made it ready for action. A rubber tube was added between the turret and the hull and inflated to keep water from entering.

The fresh-air intake utilized a wire-wrapped rubber hose, some 200 mm in diameter and 18 meters long. To its end was attached a buoy with a radio antenna. The exhaust pipes of the vehicles were fitted with pressure valves. When the tank was underwater, a sea-water system for cooling the engine was switched on. Excess water was removed by means of a sump pump. The nominal diving depth was 15 meters. The extra three meters of hose provided a safety factor.

These "diving tanks" were intended to be deployed from freight barges. From a ramp lengthened with iron rails the tank slid into the water; its course directions came to the immersed vehicle by radio from a command boat. Underwater navigation was done by means of a gyrocompass, and all the crewmen were supplied with life-saving equipment. The immersed tanks could be steered fairly easily, since the water pressure counteracted some of their weight.

In the spring of 1941, Panzer Regiment 18, which had meanwhile been transferred to Milowitz near Prague, was given orders to determine to what extent the diving tanks were still watertight and capable of diving. Instead of the long rubber snorkel, the tanks were fitted with a 3.5 meter steel pipe as an air intake, and the exhaust pipes were again fitted with non-return valves. After these tests, all the tanks were overhauled, waterproofed and prepared for diving use again. The unit was then transferred to

Eberswalde, in order to be used for diving training there.

At the beginning of June 1941, the regiment was transferred to the Bug (a river in Poland) and given the task of preparing to cross the Bug without building bridges. On June 22, 1941, at 4:45 A.M., parts of Panzer Regiment 18 crossed the Bug underwater.

A message from Feldmarschall Keitel to the OKH on July 7, 1941 stated that Hitler considered it purposeful to build future tanks with a "pre-armor" set forward from the main armor plate, strengthening them so as to offset the heightened penetrating power of anti-tank weapons. According to Hitler, the increased weight and loss of speed should be taken in the bargain. One series of "Panzer III" tanks was fitted with a pre-armor set apart from the front wall of the bow armor.

When the Russian "T-34" and "KV" tanks appeared at the start of the Russian campaign in 1941, the weakness of the German tank armament was quickly revealed to its fullest extent.

This would not have been necessary if the Army Weapons Office had followed the instructions issued personally by Hitler in 1940 and installed the 50mm KwK 39 L/60 gun, which was already available at that time, in the "Panzer III." The 37mm KwK was actually replaced only by the 50mm KwK L/42, a decision that led to unfortunate results and caused serious disagreements between Hitler and the Army Weapons Office.

On this subject, we can refer to the two original documents reproduced on page 41. They reveal very clearly the shock effect that the appearance of the modern Russian tanks had on the German military leadership.

While a total of 7992 "Panzer III" tanks was still planned for the establishment of 36 Panzer divisions as of July 17, 1941, a comment on November 29, 1941, thus after the appearance of the Russian "T-34" tank, shows an obvious change. Not only was doubt already expressed at that time as to the efficacy of the Panzer troops, but Hitler personally described the "Panzer III" tank as an unsuccessful design. As it turned out, though, this vehicle was a very progressive fighting vehicle, and with a little foresight and consideration for Guderian's wishes, this vehicle could have been the best battle tank of any in use by any of the fighting forces in 1940-41. The fact that the development of the German tanks after 1933 had been based on the concept of combat in central and western Europe, though, turned out to be, on the basis of what happened after June 22, 1941, to be a fatal error.

In 1941 the Panzerkampfwagen III, Ausf. J was introduced; it was designated "8/ZW" by Daimler-Benz and bore chassis numbers 68 001 to 69 100. The basic armor of these vehicles was strengthened from 30 to 50mm, while the short 50mm KwK L/42 was used as before.

**This picture shows a Panzer III Ausf. G of the 4th Panzer Division which was originally prepared for diving action. The watertight fittings for the bow machine gun and the primary weapon are still present.**

A N L A G E   I

Wehrwirtschaft und Ruestungsamt         Berlin, den 18. Juli 1941
Ruestungswirtschaftliche
Abteilung                               GEHEIME KOMMANDOSACHE


A k t e n v e r m e r k
ueber die Sitzung des Panzerausschusses am 17.7.1941

Beteiligt u.a. :   Chef H Ruest u. BdE       - Oberst Czimatis
                   OKH - Wa A/WuG -          - Oberst Phillips
                                             - Oberst Schroeder
                   OKH - Wa A/Wa Pruef -     - Oberst Fichtner
                   Munitionsminister         - Hauptamtsltr. Saur
                                             - Herr Hacker
                   OKW - Wi Rue Amt -        - Major Dr. Becker

Die durch die neue Fuehrerweisung befohlene Verstaerkung der Panzerwaffe macht die
Erstellung von (einschl. der bestehenden) 36 Panzerdivisionen erforderlich.
Der Vertreter des Allgemeinen Heeresamtes stellte den reinen Panzerbedarf ohne Hilfs-
und sonstige Fahrzeuge fuer diese 36 Divisionen wie folgt fest:

           Panzer II              4608 Stueck
           Panzer III             7992 Stueck
           Panzer IV              2160 Stueck
           Befehlswagen            684 Stueck

                                 15444 Stueck Panzerfahrzeuge.

Um einen Vergleichsmasstab zu bekommen, wurde mitgeteilt, dass beim Feldheer zur Zeit
4000 Fahrzeuge dieser Art vorhanden sind.
Oberst Phillips als Vertreter des Waffenamtes forderte zur Erstellung und Aufrecht-
erhaltung dieser Divisionen den Ausbau der Panzerfertigungsstaetten auf 2000 Einheiten
monatlich. Zur Zeit sind etwa 600 Einheiten erreicht.
Dass der Ausbau von 2000 Einheiten nicht verwirklicht werden kann, ist schon daraus
zu ersehen, dass der Ausbau der derzeitigen Stufe des Panzerprogramms mit 5500 Werkzeugmaschinen in Rueckstand ist, von denen monatlich 2 bis 300 geliefert werden.
Man einigte sich nun zunaechst auf den Ausbau von 900 Einheiten als minimum, fuer den
ueber den Rueckstand von 5500 noch 12 bis 15000 Werkzeugmaschinen gefordert werden
muessen.

- 2 -

Dr. Porsche fuehrte aus, dass durch eine Vereinfachung der Konstruktion eine Ermaessigung des Maschinenbedarfs von 15 bis 20% moeglich sei.

Der Vertreter des Munitionsministers brachte zum Ausdruck, dass es der Wunsch des Ministers sei, statt der handwerksmaessigen Fertigung mindestens ein Montagewerk fuer Fliessarbeit einzurichten. Mit Ruecksicht auf die Luftlage schlaegt er als Standort Schlesien oder die Ostmark (Hallein) vor.

Saemtliche Vertreter der Industrie widersprachen, Werke ohne Anlehnung an bestehende Fertigungsstaetten zu erstellen, denn bei neu- und selbststaendig anlaufenden Werken sei mit einem brauchbaren Ausstoss vor mehreren Jahren - wie die Erfahrung gezeigt habe - nicht zu rechnen.

Es wurde beschlossen, dass fuer die Neuplanung die Arbeitsvorbereitung bei der MAN liegen soll. Als Vorbereiter sollten Angestellte zur Verfuegung stellen das Volkswagenwerk, Opel, Ford, Henschel, Daimler und Auto-Union.

Es wurde weiter beschlossen auch die Ausstattung der 18 Divisionen mot, welche neben den 36 Panzerdivisionen gehalten werden sollen, in die Bearbeitung des Panzerausschusses zu nehmen, was die Vertreter des Heeres sehr begruessten.

Im Zuge dieser Aussprache wurde zum Ausdruck gebracht, dass das Allgemeine Heeresamt damit rechnet, dass bei den im Osten eingesetzten Panzern im Endeffekt zur Zeit mit einem Gesamtausfall von 50% gerechnet werde.

Fuer den im Zuge der Planung durchzufuehrenden Ausbau von 900 Einheiten wurden als bevorzugt fertigzustellen bezeichnet: Fertigungsstaetten fuer die Leistung von monatlich:

    1.) 1000 Motoren                         (in Anlehnung an Maybach)
    2.) 600 Getrieben
    3.) 1200 Satz Panzerwagenketten
    4.) 2250 Satz Zugkraftwagenketten
    5.) 500 Blechen                          (in Anlehnung an Alpine Montan).

Daneben sollen 2 Montagewerke geplant werden, eines in Anlehnung an MAN fuer 150 Einheiten und eines in Anlehnung an Deutsche Eisenwerke Muelheim fuer 150 Einheiten. Damit wuerde der derzeitige Stand von 600 Einheiten auf die geplanten 900 erhoeht.

Es sei auch vorgesehen, dass ein Panzerwerk durch Roechling errichtet werden soll.

Im Zuge der Aussprache ueber diese neuen Werke wurde erwaehnt, dass allein beim Ausbau der Panzerwagenfertigung ein Bauueberhang, also Rueckstand, von 85 000 000 RM vorhanden sei. Fuer das 3. am 1.9. beginnende Kriegswirtschaftsjahr kaemen im Zuge dieser Neuplanung noch 100 000 000 RM dazu, sodass zusammen mit 200 Millionen RM Bausumme zu rechnen sei.

Oberst Phillips bemerkte dazu, dass dem Waffenamt bisher ein Bauvolumen von 900 Millionen zugestanden gewesen sei, welches jedoch auf 450 000 000 gekuerzt werden solle. In diesen 450 Millionen solle auch noch der Schnellplan eingeschlossen sein.

Diese Zahlen kennzeichnen die Schwierigkeiten, die der Durchfuehrung der Planung entgegenstehen.

Es gab dann Direktor Oberlaender von Daimler einen Ueberblick ueber die Ersatzteillage. Seit der letzten Sitzung seien 452 t Ersatzteile weiter abgeliefert worden. 179 t liegen versandbereit und weitere betraechtliche Mengen wuerden taeglich fertiggestellt.

Schliesslich hielt noch Oberst Fichtner von Wa Pruef 6 einen Vortrag ueber russische Panzer. Als Termin fuer die naechste Tagung des Panzerausschusses wurde der

<u>31.Juli 1941 - 10 Uhr vormittags</u>

in Kummersdorf festgesetzt. Es sollen bei dieser Gelegenheit russische Panzer, die wieder fahrbereit gemacht worden sind, vorgefuehrt werden.

A N L A G E   II

GEHEIME KOMMANDOSACHE

B e s p r e c h u n g

am 29. November 1941, 12:00 - 14:00 Uhr

Ort: Reichskanzlei

Teilnehmer:

    Der   F u e h r e r

OKW: Generalfeldmarschall Keitel

    General der Artillerie Jodel

    Oberst d.G. Schwandt

OKH: Generalfeldmarschall von Brauchitsch

    General der Artillerie von Leb, Chef Waffenamt

    Oberst d.G. Koehler, Chef des Stabes beim Chef H.R.

    Oberst Phillips W. u. G. Wa.

    Oberst Fichtner Wa., Pruef. 6

    Oberst I.G. von Gyldenfeldt, 1. Genst. Offizier beim Ob.d.H.

    Oberst d.G. Loehr, Chef des Stabes Wa.

Ministerium fuer Bewaffnung und Munition:

    Minister Dr. Todt

    Direktor Schaedte

    Oberdienstleiter Saur

Industrie:

    Professor Dr. Porsche

    Direktor Hacker, Steyrwerke

    Direktor Dr. Rohland, Vereinigte Stahlwerke

Grund der Besprechung:

    Panzerwagenfertigung und Panzerabwehr.

Der Fuehrer:

Die Erfolge der letzten Kriegsjahre sind neben der Tapferkeit der Soldaten und der Kunst der Fuehrung, zwei Faktoren zu verdanken: Der Luftwaffe und der Panzerwaffe. Heute ist die Panzerwaffe zu behandeln.

Panzer-Divisionen taten zunaechst den entscheidenden Stoss nach Polen. Zehn Panzer-Divisionen fuehrten den entscheidenden Durchbruch im Westen bis zur Kanalkueste und fuehrten die Vernichtung der Westheere herbei. Eine Panzer-Division wendete zunaechst die Krise in Afrika. Zwanzig Panzer-Divisionen ermoeglichten die Erfolge im Osten. Die Erfahrungen im Ostfeldzug beweisen, dass wir an einem Wendepunkt stehen.

Unsere Panzerabwehrwaffen sind einem Teil der Russen Panzer und dem englischen Infanterie-Panzer nicht mehr voll gewachsen. Die Panzerstaerken unserer Kampfwagen genuegen den russischen Panzerabwehrwaffen gegenueber nicht mehr.

Das Zeitalter des Panzers kann bald vorueber sein. Es kommt fuer uns darauf an, die Zeit fuer die Erfuellung unserer Aufgaben noch zu nutzen in der die Panzerverbaende noch als Angriffswaffe verwendet werden koennen. Sind unsere europaeischen Aufgaben erfuellt, so kann uns die Entwicklung nur recht sein. Denn wir werden dann in der Verteidigung des Erbes, den Sieg der Abwehrkraft ueber dem Panzer ausnuetzend, uns jeden Angriffs erwehren koennen.

Um jedoch unsere letzten Ziele mit dem Panzer ereichen zu koennen, sind wir verpflichtet, die Panzerwaffe und die Abwehrwaffen den Feindstaaten gegenueber widerstandsfaehig, moeglichst ueberlegen zu machen.

Hierzu ist notwendig:

1. Es muss davon ausgegangen werden, dass die Abwehrkraft gegen die Kampfwagen sich steigert und der Gegner die Panzergranate 40 kennt. Die Aufkaeufe an Wolffram auf dem Weltmarkt spricht davon, dass der Gegner seinerseits zur Herstellung solcher Panzergranaten uebergegangen ist oder uebergeht. Ob der Feind Kenntnis von unseren Hohlgeschossen hat ist eine offene Frage. Es ist eine geschichtliche Erfahrung, dass derjenige, der jahrelang eine Entwicklung betreibt wie wir, allmaehlich Scheuklappen bekommt; und dass derjenige, dem die Entwicklung als fertiges Produkt in die Haende faellt, nicht nur das Produkt als solches uebernimmt sondern schneller zu Verbesserungen kommt. Das Beispiel hierfuer ist der Russe selbst der aus den Ausbeuten der Produkte anderer Nationen, erheblichen Nutzen gezogen hat.

   Der Uebergang zu hoeherem Kaliber der Pak in allen Staaten, zwingt zur Erhoehung der Panzerstaerken.

   Da schlecht gepanzerte Kampfwagen unbrauchbar sind, muss die Verstaerkung der Panzer selbst auf Kosten der Schnelligkeit vorgenommen werden.

   Ob eine sture Erhoehung der Panzerstaerke vorgenommen oder zum Vorpanzer uebergegangen werden soll, ist schwer zu entscheiden. Fuer die Panzergranate 40 ist die Verstaerkung der Panzerung durch den Vorpanzer geloest. Ein Teil der feindlichen Pak wird durch den Vorpanzer ausgeschaltet. Nicht geloest wird durch den Vorpanzer der Schutz vor dem Beschuss mit Hohlgranaten. Beim ersten Treffer wird der Vorpanzer zertruemmert. Der Vorpanzer ist also hier nur ein Teilschutz. Erste Forderung ist, bis zum Fruehjahr 1942 die Kampfwagen aller Einheiten mit dem Vorpanzer zu versehen. Wo die Sehschlitze den Vorpanzer nicht gestatten, muss die Panzerstaerke selbst erhoeht werden.

2. Die Durchschlagskraft der eigenen Panzerabwehr muss erhoeht werden. Eine Ver-

zoegerung der Produktion an Panzerabwehrwaffen darf dadurch nicht eintreten.

Es kann kein Zweifel daran bestehen, dass die Gegner hinsichtlich ihrer Panzerabwehr, Verbesserungen vornehmen werden.

Die Modelle die zur Zeit an der Front verwandt werden, muessen ebenso wie die neu zu produzierenden Modelle bei Steigerung der Schussentfernung, eine groessere Durchschlagsleistung bekommen.

Der Fuehrer spricht von Neukonstruktionen:

Die Fronterfahrung ergibt ein Konglomerat verschiedenster Typen. Typenbegrenzung ist erstes Erfordernis. Der Panzer III ist eine nicht gelungene Konstruktion. Die Schwere des Wagens steht in keinem Verhaeltnis zu der unzulaenglichen Bewaffnung mit der 3,7 und 5 cm KwK.

Es ist notwendig, sich auf drei Typen zu beschraenken:

    Leichte Type als Aufklaerungsfahrzeug: Der bisherige Panzer III

    Mittlere Type: Panzer IV

    Schwere Type: (Konstruktion Henschel-Waffenamt und Porsche).

Dazu ist eine ueberschwere Type durch Professor Porsche zu entwickeln, damit wir hier in der Lage sind fuer die Zukunft vorzuhalten. Hohe soldatische Werte verpflichten auch technische Hochwerte zu liefern.

Unabhaengig davon verlaeuft der Bau der Sturmgeschuetze.

Ziel muss sein, den motorisierten Divisionen durch Zuteilung von Kampfwagen die Panzerspitze zu geben und durch sie die z.Zt. dazu benutzten Sturmgeschuetze zu ersetzen. Die Sturmwagen gehoeren zur Infanterie, die sie unmittelbar begleiten muessen. Die Sturmwagen muessen ein Geschuetz erhalten das hohe Wirksamkeit gegen Panzer besitzt. Die Sturmgeschuetze muessen mehr Munition bekommen.

Die Schnelligkeit des Sturmgeschuetzes kann herabgesetzt werden, wenn es lediglich als Infanterie-Begleitschutz verwandt wird und ihm Aufgaben die der Panzerwaffe zufallen, abgenommen werden.

Fuer die Verwendung bei Infanterie-Divisionen genuegt eine Geschwindigkeit von 12 km in der Stunde.

Zur Produktion uebergehend fuehrt der Fuehrer aus:

Es ist ein bitterer Irrtum, zu glauben, dass wir in der Ausschoepfung fabrikatorischer Moeglichkeiten ein fuehrender Staat sind.

Die Konstruktionen stehen im Gegensatz zu den praktischen Produktionsmoeglichkeiten z.B. in der Ausschoepfung der Moeglichkeiten des Stanzens und Giessens usw.

Es muss also viel weitgehender bei den Konstruktionen auf die Produktion Ruecksicht genommen werden.

Es empfiehlt sich hierzu vornehmlich die Privatindustrie beratend heranzuziehen. Sie ist gezwungen, zu kalkulieren und oekonomisch zu arbeiten, waehrend Staatsbetriebe im

- 4 -

Allgemeinen aus dem Vollen arbeiten, da ihre Beamten und Angestellten durch den Staat gehalten werden.

Durch Vereinfachung der Konstruktionselemente muss die fluessige Massenproduktion gefoerdert werden.

Es ist keinesfalls zutreffend, anzunehmen, dass dadurch der Zweck oder die Brauchbarkeit der Waffe vernachlaessigt wird. Die Normisierung und Typisierung beguenstigt das Ausschlachten nicht mehr einsatzfaehiger Fahrzeuge. Wir schaffen dadurch ein natuerliches Ersatzteillager, das uns z.Zt. bei der Verschiedenartigkeit der Typen in hohem Masse allenthalben fehlt. Die Teile eines Fahrzeuges muessen mit denen eines anderen jederzeit ausgewechselt werden koennen z.B. die Laufrollen.

Es ist nicht zu vertreten, dass wir heute Fahrzeuge bauen, die 120 Jahre ueberdauern koennen, waehrend wir ebenso wissen, dass sie nach 2 oder 3 Jahren ueberlebt sind.

Die Zugmaschinen muessen Typen einfachster Art werden. Es bedarf keiner Ruecksicht auf Schoenheit, es kommt vielmehr auf Zweckmaessigkeit an die auch eine Massenproduktion erlaubt. Im Osten werden wir mit heuschreckenartigen Fahrzeugen gute Erfahrungen machen.

Dass man mit Nutzen zu Vereinfachungen uebergehen kann, zeigt die Konstruktion eines neuen Maschinengewehres, das bei Vereinfachung der Konstruktion, Verringerung der Kosten, eine bessere Wirkung hat und feldbrauchbarer ist.

Ein Vorhalten bei neuen Konstruktionen ist notwendig, wenn wir in unseren Konstruktionen noch 1 bis 2 Jahre auf der Hoehe durchhalten, dann bin ich zufrieden. In der Abwehr werden wir dann spaeter mit dem Hohlgeschoss allen Angriffswaffen ueberlegen sein.

Der Fuehrer spricht sodann von der notwendigen Verkuerzung der Transportwege im Kreislauf des Materials, der heute unuebersehbar geworden ist. Geringerer Materialabfall bei der Fertigung ist anzustreben.

Das Fehlen von notwendigen Rohstoffen darf nicht dazu fuehren, auf schlechteres Ersatzmaterial zu verzichten und die Konstruktion aufzugeben. Dem Soldaten an der Front ist eine Waffe lieber die eine geringere Leistung taetigt, als keine Waffe zu haben.

D e r   S c h i c k s a l s k a m p f   d e s   d e u t s c h e n   V o l k e s   d a r f   n i c h t   d a d u r c h   i n   F r a g e   g e s t e l l t   w e r d e n ,   d a s s   d i e   F a n t a s i e   K r i e g s m a s c h i n e n   e r d e n k t ,   d i e   n i c h t   i n   M a s s e n   p r o d u z i e r t   w e r d e n   k o e n n e n .

Der Fuehrer betont nach Aussprachen, waehrend deren Minister Dr. Todt, Oberst Fichtner und weitere Herren zu Worte kamen, dass der Deutsche Soldat die besten Waffen verdient und es eine Kapitulation vor dem Fehlen technischer Moeglichkeiten nicht gibt.

    Aus der Diskussion ist erwaehnenswert:

Minister Dr. Todt versichert, dass durch Vereinfachungen 30% Arbeiter fuer die Produktion eingespart werden koennen.

Oberst Fichtner gibt als Auffassung der Front bekannt, dass hoehere Schussentfernungen der Geschuetze, der Truppe lieber ist, als staerkere Panzer.

                                                                          gez. Schmidt
                                                                          7. 12. 1941

This interior view of the "ZW" vehicle shows the driver's seat, steering lever and vision port. The gyrocompass can be seen at the upper left. At the right, beside the driver's seat, is the gearbox with the shift lever, and above it is the instrument panel. At the left in the background is the steering brake; the external drive sprocket is attached to its far end.

This close-up view of the driver's area shows the driver's optics when folded in, with the gyrocompass beside them. In the left sidewall there is another viewing port. A gas mask and machine-gun ammunition sacks are attached to the sidewall.

The radio-operator's seat next to that of the driver shows the bow machine gun with its aiming optics and the hanging machine-gun ammunition sacks. At the radio-operator's left, above the equipment, is the radio.

This drawing shows the design of the "KFF 2" driver's optics in connection with the Fahrersehklappe 50 (Driver's Vision Hatch 50). The upper part of the drawing shows its use while on the march, the lower part that used during combat.

1. Head pad
2. Head support
3. Eyepieces
4. Periscopes
5. Boreholes for driver's periscope
6. Adjusting lever
7. Housing
8. Armor plate (a, open, b, closed)
9. Sliding rod
10. Nose protector
11. Forehead protector
12. Protective glass
13. Circle segment

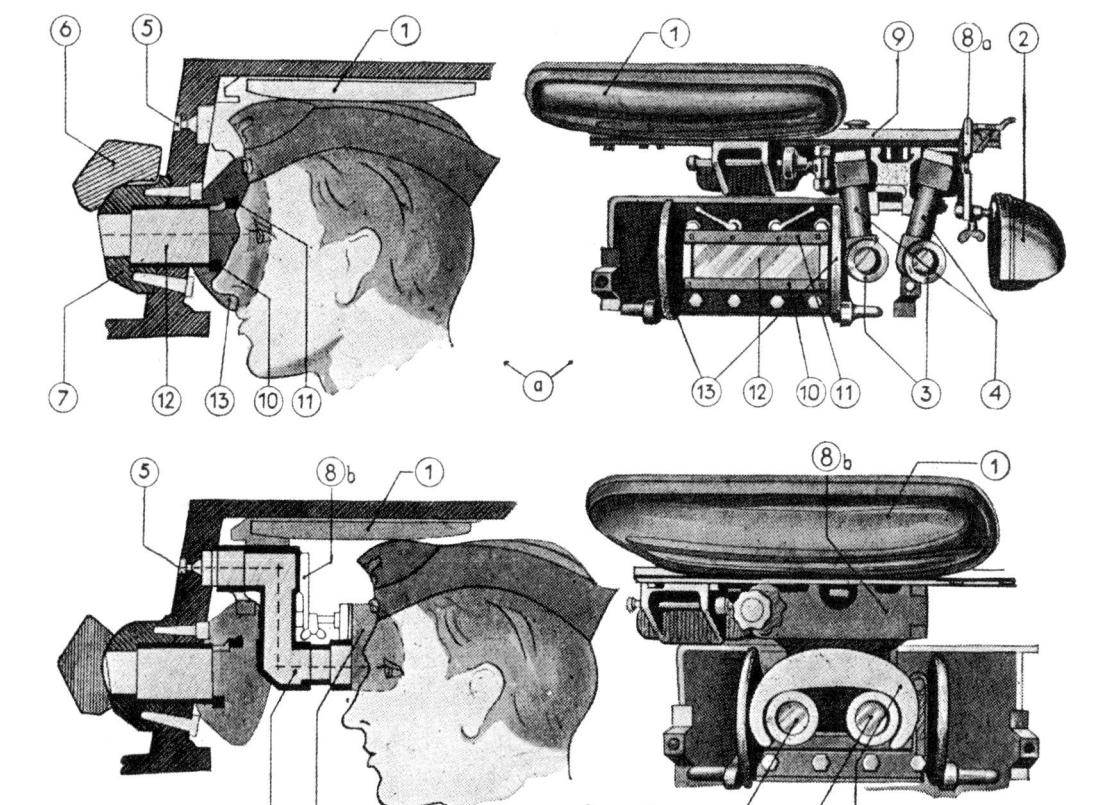

This interior view of the fighting compartment shows the back wall of the turret with the commander's cupola and the commander's and gunner's seats.

Track link

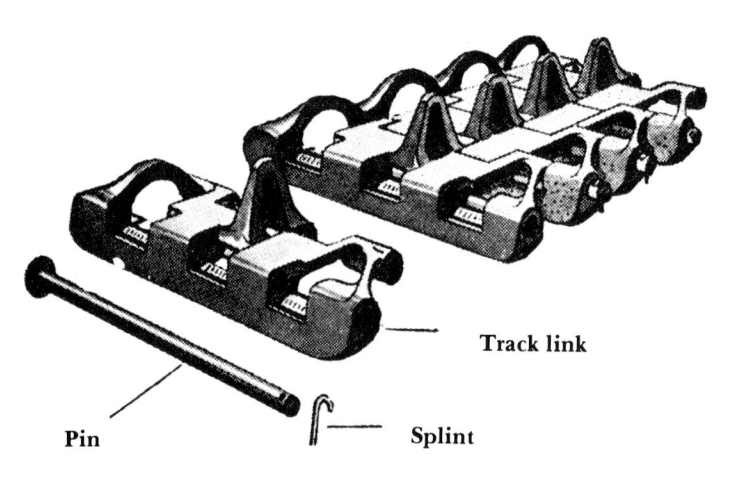

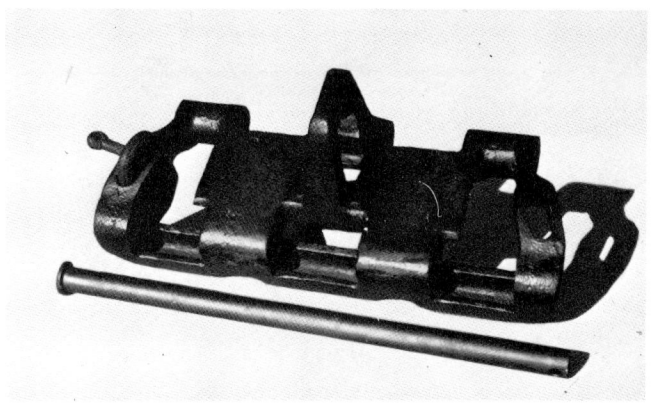

Pin    Splint

**As of Ausf. H, the track was widened from 380 to 400 mm. This drawing shows the construction of the unlubricated Kgs 61/400/120 type track. The picture complements the drawing.**

After the 50mm KwK was introduced, two blue warning lights were mounted on the front wall of the bow armor in front of the tank driver. They served to inform the driver if, while the tank was underway, the 50mm KwK barrel extended over the outside limits of the vehicle when the turret was turned. The warning lights were switched on by a switch that was screwed onto the top of the bow armor in Ausf. E and F and built into the release-ring carrier of Ausf. G and subsequent types.

The introduction of 50 mm armor plate resulted in the use of a Fahrersehklappe 50 (Driver's Visor Flap 50), consisting of a housing screwed onto the front wall of the bow armor and covered by a sliding panel.

From the Ausf. J on, only the "KFF 2" driver's optics could be used. The "Kugelblende 50" ball mantelet was also used from Ausf. J on.

The racks and containers for the engine and weapons equipment and supplies carried in the upper part of the superstructure showed differences as to their location and the number of them used in types E. F. G, H and J.

Outside, screwed onto the back wall of the turret (and retro-fitted to all Panzerkampfwagen III's), was a box that held the crew's packs.

In the bow armor of the hull there were, through Ausf. H, two divided, boltable entry hatches for the driver and radio-operator. Beginning with Ausf. J, there were only two one-piece servicing hatches. Through Ausf. H, there

**For driving on ice and snow, snow grippers of various designs could be attached to the tracks. If necessary, track widening pieces were available. Sometimes track links were also seen with welded-on raised areas.**

50

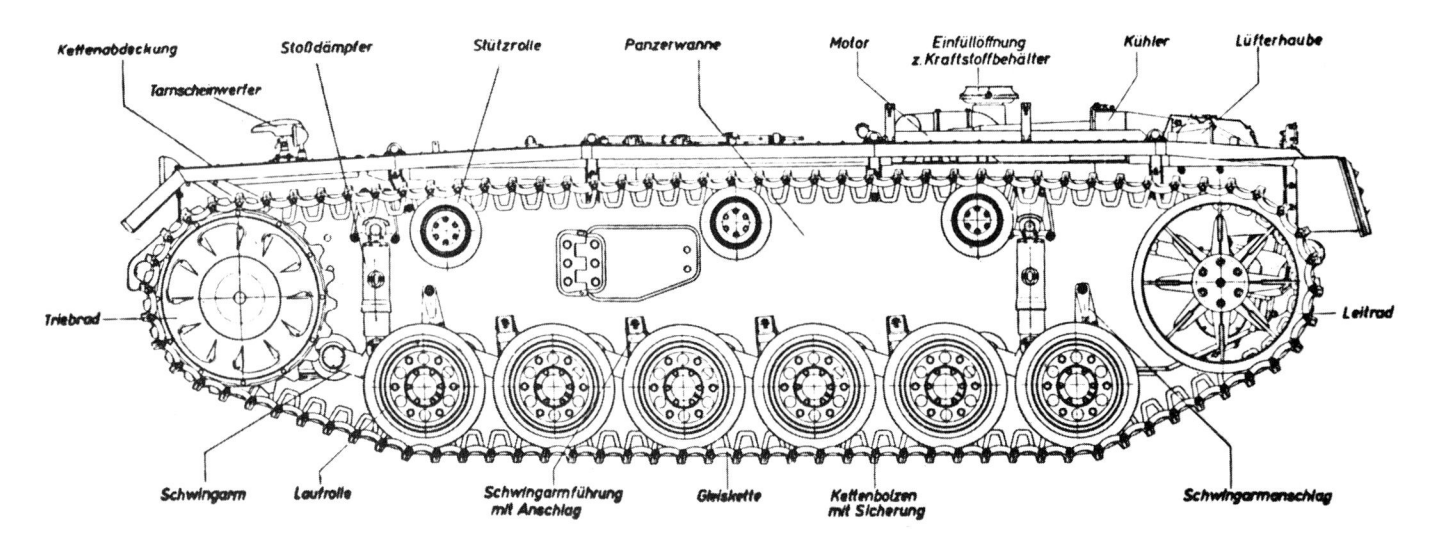

As of Ausf. J, the basic armor of Panzerkampfwagen III was increased from 30 to 50 mm. This drawing shows the chassis of Ausf. J to L.

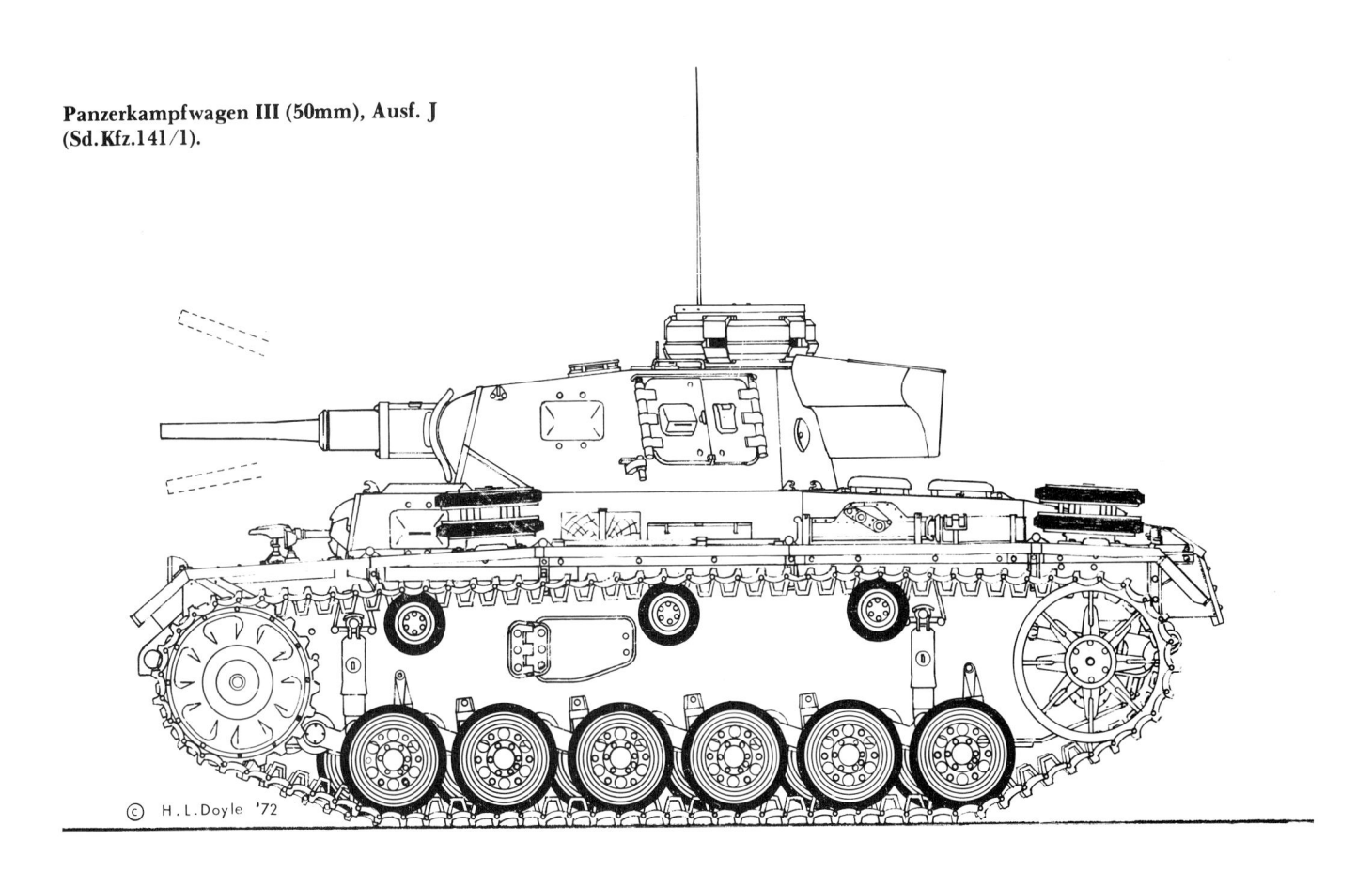

Panzerkampfwagen III (50mm), Ausf. J (Sd.Kfz.141/1).

The first series of Ausf. J was still fitted with the short 50mm KwK L/42 gun.

On April 4, 1942 Hitler was shown the Panzer III with the 50mm Pak 38 installed. This weapon had a muzzle brake which was not used on the KwK version.

were two towing-hook attachments with bolts on each end of the tank's hull, to allow the vehicle to be towed, but beginning with Ausf. J there were only two welded-on eyes. Delbag air filters were used through Ausf. H; they were attached to the engine-compartment side of the bulkhead and linked with the carburetor by a forked tube. Beginning with Ausf. J, Mahle centrifugal oil filters located in the engine compartment were used; they were connected mechanically to the steering and support brakes, whereas through Ausf. K (Armored Command Vehicle) the connection had been hydraulic.

According to a message from the authorizing General in charge of motor vehicles, dated July 21, 1941, additional capacities and production facilities for expanded tank production were being set up. Among them were a Daimler-Benz AG factory, the Fahrzeugwerke Friedrich Krupp AG, Fross-Büssing in Vienna, Tatra Wagenwerk in Kolin, Framo in Hainichen, and branches of MAN, Henschel, Hanomag, Auto-Union and NSU.

In terms of quantities, there were 327 Panzer III tanks with the 37mm KwK gun and 1174 with the 50mm KwK in service with the army in the field as of July 1, 1941. The number of tanks with the 37mm gun has decreased to 131 by April 1, 1942, while the number of tanks with the 50mm gun had risen to 1893. The monthly production of "Panzer III" tanks for January 1942 had been set at 190 units, but in fact, what with transport difficulties involving guns and armor plate, only 159 could be delivered.

At the end of 1941 the second series of Panzerkampfwagen III, Ausf. J, with chassis numbers from 72 001 to 74 100, was delivered. This series finally had the 50mm KwK L/60 gun as standard equipment. A comparison of this weapon, supplied by both the Karges-Hammer firm in Braunschweig and that of Franz Garny in Frankfurt am Main, shows the following improvements:

A column of Panzerkampfwagen III Ausf. J on the march. The storage boxes added to the back of the turret were typical of Panzer III and IV.

A Panzer III Ausf. J of the 14th Panzer Division during service in Russia. The commander and gunner are observing.

|  | 50mm KwK 39 L/42 | 50mm KwK 39 L/60 |
|---|---|---|
| Barrel length | 2103 mm = L/42 | 3000 mm = L/60 |
| Maximum velocity | | |
|   Sprenggranate | 450 meters/second | 550 meters/second |
|   Panzergranate | 685 meters/second | 835 meters/second |
|   Panzergranate 40 | Not available | 1180 meters/second |
| Shell weight | | |
|   Explosive shell | 1.85 kp | 1.86 kp |
|   Antitank shell | 2.06 kp | 2.06 kp |
|   Antitank shell 40 | not available | 0.92 kp |
| Gross weight installed | 400 kp | 435 kp |

**The second series of Ausf. J finally had the long 50mm KwK 39 L/60. During the course of Ausf. J production, the vision ports and flaps in the sides of the turrets were eliminated.**

The ammunition supply for these vehicles, now referred to as "Sd.Kfz.141/1", was cut from 99 to 84 shells, while the supply of 3750 machine-gun rounds remained unchanged. Until the installation of the long 50mm tank gun, all Panzer III tanks had a folding seat for the loader attached to the bulkhead that separated the fighting and engine compartments. This seat was then eliminated. During the course of Ausf. J production too, the vision ports and their flaps built into the right and left sidewalls of the turret, and the port flap to the right of the mantelet, were eliminated.

Forty production vehicles with the 50mm KwK L/60 were delivered by the end of 1941.

The troops had never been happy about the elimination of the tank gun from the Befehlspanzer (armored command vehicle). There were very often situations in tank battles that could be mastered only through personal examples set by the commander. To do this, he needed a fully armed vehicle. It had also proved to be disadvantageous to use tanks of a special type as command vehicles. If they were put out of action, there were immediate command difficulties.

Therefore Daimler-Benz received a development contract for a new armored command tank in January of 1941. It was to carry the 50mm KwK 39 as its primary armament and be reequipped with means available to the troops. Thus it had to be possible to install additional radio sets with no trouble. Thus was the Panzerbefehlswagen III, Ausf. K created (called "8a/ZW" by Daimler-Benz, and with chassis numbers 70 201 to 70 400). The vehicle began troop testing in August of 1941; it was similar to the "8/ZW" tank type. Its fighting weight was some 23 tons, its crew numbered five men, with the loader also serving as a second radioman. The price of the vehicle (without weapons) was 110,000 Reichsmark. Series production lasted from August 1942 to August 1943.

The radio equipment of German armored vehicles was given particular attention since the beginning of their development. The Panzerkampfwagen III was equipped with one ultra-short-wave transmitter and two ultra-short-wave receivers. The transmitter was located in front of the radioman, the two receivers to his left. The rod antenna was attached outside on the right side of the

**These pictures show a vehicle of the 9th series, Panzer III, Ausf. L. It was produced regularly with a set-off pre-armor 20 mm thick before the driver's front wall and the gun mount.**

vehicle and could be folded backward and placed in a wooden protector when not in use. The antenna itself, a round hollow rod made of thin, hardened sheet copper, was two meters long and activated from inside the vehicle.

The complete Panzer III production in 1941 numbered 1845 units.

At the beginning of 1942 the delivery of Panzerkampfwagen III (50mm) Ausf. L (Sd.Kfz.141/1) began; its Daimler-Benz designation was "9/ZW", its chassis numbers ran from 74 101 to 76 000. Only in production form did this vehicle have a separate pre-armor plate, 20

**Panzerkampfwagen III (50mm), Ausf. J (Sd.Kfz.141).**

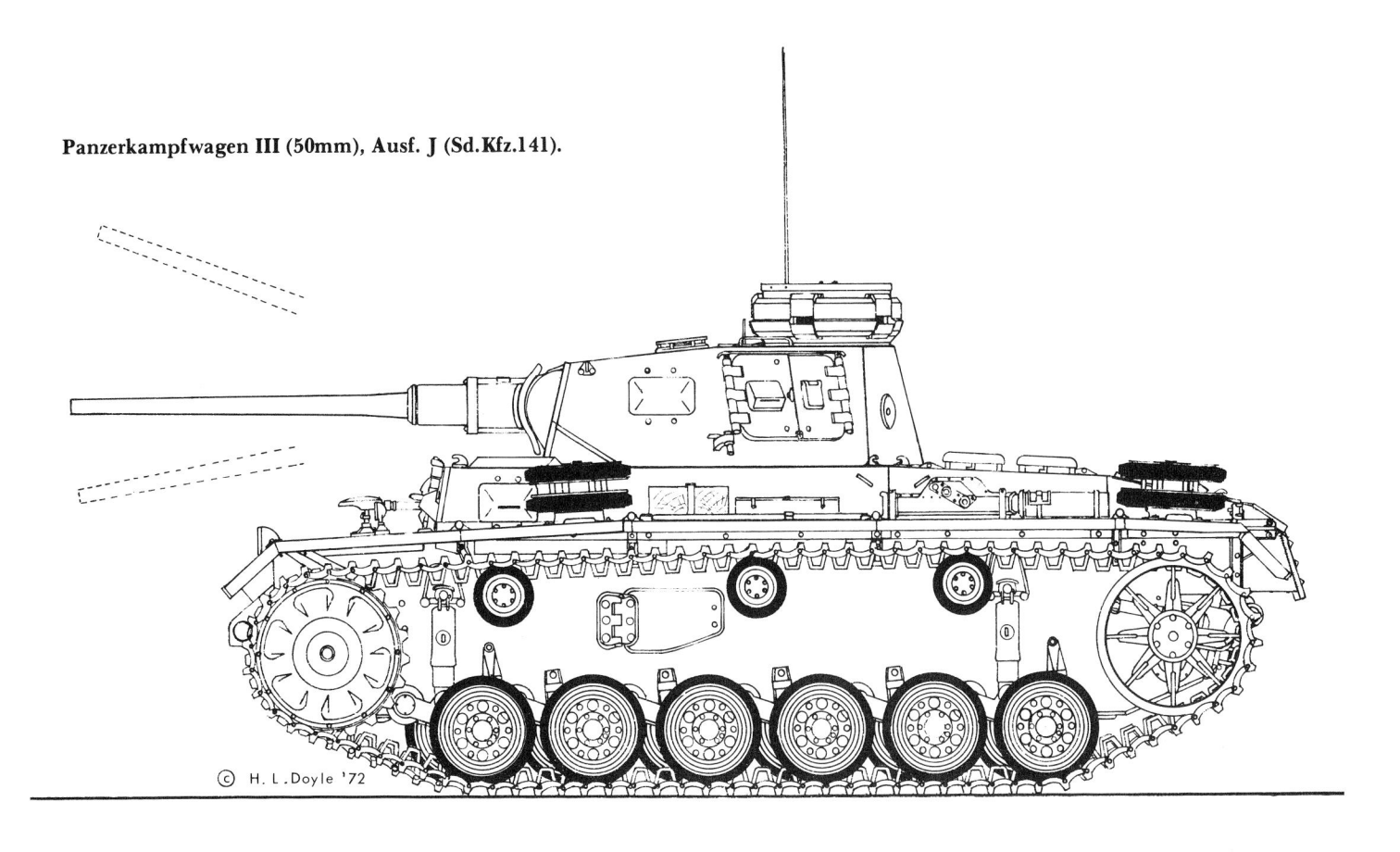

© H. L. Doyle '72

**Panzer III Ausf. J tanks ready to leave the Maschinenfabrik Augsburg-Nürnberg.**

mm thick, in front of the driver's front wall and the gun mantelet. Until Ausf. H, the weight of the barrel and its cradle at the mantelet in front was counteracted by a balance weight in back on the deflector of the gun cradle. Starting with Type J, this was accomplished in the mantelet by a spring equalizer that was suspended from the right side of the mantelet and the carrier ring by one hook screwed onto each. When pre-armor was used, its pressure at the front was equalized by another spring equalizer. But a number of these vehicles had to be delivered to the troops without it because of a shortage of additional armor plate. The fighting weight of the Ausf. L was 22.3 tons. During the production run of this version, the emergency exit hatches on both sides of the hull, used since Ausf. E, were eliminated. Vehicles fitted for tropical service now took in combustion air through the fighting compartment, and curved heat deflectors were attached to the exhaust system.

On the basis of experience gained during the Russian winter, a cold water exchanger was added to make possible the exchange of warmed cooling liquid from one vehicle to another.

While the 50mm KwK L/42 was normally supplied with 59 explosive shells, 36 antitank shells and five optional shells, the supply for the L/60 gun was 84 shells. For the two machine guns, 25 belts with 150 rounds each — 3750 rounds in all — were available. By eliminating the hull exit hatches, the ammunition supply for the tank gun could be raised to 98 shells. In addition, 24 signal rounds were also carried.

At this time efforts were also made to install the Waffe 0725 weapon with a conical barrel in the Panzer III tank. Production of this effective armor-piercing weapon could not be initiated because of the shortage of materials (molybdenum steel).

In March of 1942 the reequipping of the Panzer III with the rotating turret of the Panzer IV was debated. With concern for the resulting problems, such as strongly increased gross weight, new running gear, new tracks and weight distribution. The Panzer III was to continue in production with the 50mm KwK L/60, but production of the Panther tank was to replace the Panzer III as soon as possible, since the Panzer III no longer possessed its old fighting value. Hitler considered it possible in view of the favorable production conditions for Panzer III and Panzer IV to let production of the former continue at full capacity.

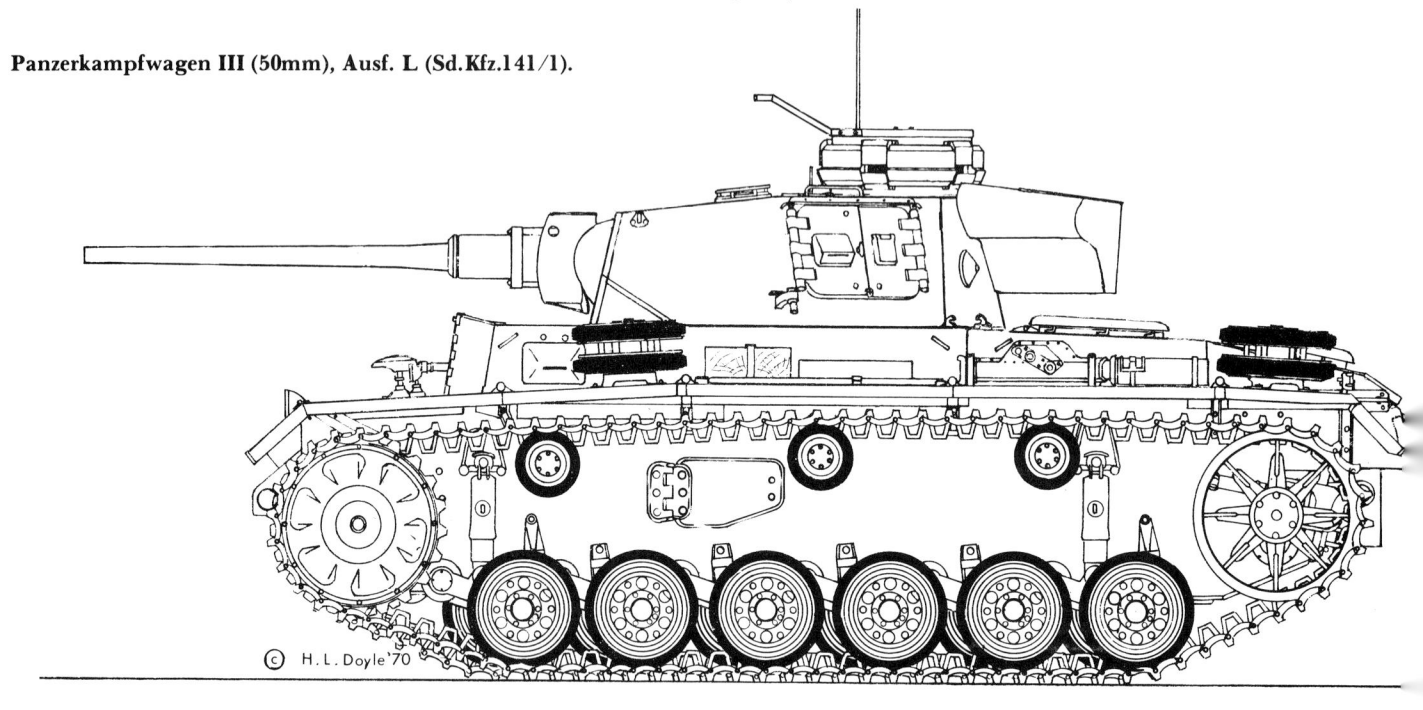

**Panzerkampfwagen III (50mm), Ausf. L (Sd.Kfz.141/1).**

**During production of the Ausf. L, the emergency exit hatches located in both sides of the hull since Type E were eliminated.**

On April 4, 1942 Hitler was shown the Panzer III with the 50mm Pak 38 gun.

The gross weight of the Ausf. M series of the Panzer III, which began in 1942, was 21,130 kp. Its Daimler-Benz designation was "10/ZW", its chassis numbers from 76 001 to 78 000. A basic change in the exhaust system made the vehicle capable of wading to a depth of 1.3 meters. The price of this version (without armaments) was 96,183 Reichsmark per vehicle.

On June 4, 1942 Hitler declared that he was in agreement with Speer's suggestion that the production of the 75mm L/24 gun should not begin anew. On the other hand, the existing 450 guns should be installed into the current type of Panzer III turret in July, August and September of 1942, making a unique series. The still remaining 150 barrels of this weapon could also be fitted with breeches and bases and be available for installation in Panzer III tanks.

At the end of November 1942, the first front-line reports on the installation of this weapon in the Panzer III tank were available. On the basis of positive results, they led to Hitler's instructions that the available barrels were first to be fitted with breeches, and that resumption of barrel production for the short 75mm weapon could be envisioned. This would mean that all subsequently produced Panzer III tanks, as well as a goodly number of the tanks on hand to be rebuilt for the 50mm L/42 gun, could be equipped with the 75mm L/24 gun instead.

At the end of 1942 Hitler ordered an investigation of the extent to which cessation of Panzer III production could be balanced by a corresponding increase in Sturmgeschütz (assault gun) production. On the basis of the reported possibility for switching production, Hitler decided on December 1, 1942 that, instead of the Panzer III, production of the Sturmgeschütz should be expanded beyond previous plans, from 120 units in December 1942 to 220 units by July of 1943.

In 1942, 2555 "ZW" vehicles were completed. Of the Ausf. M tanks, the MIAG firm's Amme works in Braunschweig-Neupetrietor delivered 100 vehicles, without primary armament, to the Waggonfabrik Wegmann AG in Kassel-Rothenditmold for conversion to flame-throwing tanks intended for use at Stalingrad. On January 18, 1943 Speer had to inform Hitler that, on the basis of limitations and the related production difficulties with these Panzerkampfwagen III (Fl) (Sd.Kfz.141/3), no final production figures could be given. Nevertheless, efforts would soon be made to attain maximum production. Instead of the normal 50mm primary armament, a flame-throwing barrel with a 14 mm jet was installed. To power the pumping system of the flamethrower, a DKW two-stroke motor was used. The burning oil supply of 1000 liters made possible the emission of 70 to 80 jets of flame,

The crew practices leaving and Ausf. G tank modified for submerging. The two machine guns could be removed quite easily for use in close-range defense outside the vehicle.

each lasting two to three seconds. The jet of flame reached 55 to 60 meters. But these vehicles, which were first issued to two companies in December of 1942, were a disappointment, as the jet of flame actually extended only 35 meters. The vehicle's fighting weight was about 23 tons. With a crew of three men, the radio equipment consisted of one Fu 5 and one Fu 2. On the basis of combat experience, these vehicles were never sent to their original scene of action, and most of them were soon converted back to gun-carrying tanks. In November of 1944 Hitler returned to the subject of these vehicles and requested investigation as to the extent to which available flamethrowing tanks or assault-gun vehicles could be re-equipped with flamethrowers in a very short time. On February 12, 1944 he had already ordered that a large number of flamethrowing tanks (at least 20 to 30 of them) had to be prepared for a special action.

On December 3, 1944 Speer reported after consultation that as far as could be determined, the ordered flamethrowing tank action was ready to commence with a total

The 10th series of Panzer III, Ausf. M were given a basic change in the form of a modified exhaust system. With it, the vehicle was capable of wading through water up to 1.3 meters deep.

of 35 tanks. No further details are known.

At the beginning of January 1943 it was reported to Hitler that, instead of the Panzer III with the 50mm KwK L/60, production from January 1 to May 12, 1943 was planned for 535 units, divided into the following numbers of different types on the basis of an order to switch production:

a. 235 additional Sturmgeschütz with the 75mm StuK L/48,
b. the 100 flamethrowing tanks already mentioned,
c. 56 Panzer III tanks with 50mm L/60 for Turkey, and
d. 144 Panzer III tanks with the 75mm L/24 gun.

**Panzerkampfwagen III (50mm), Type M (Sd.Kfz.141/1).**

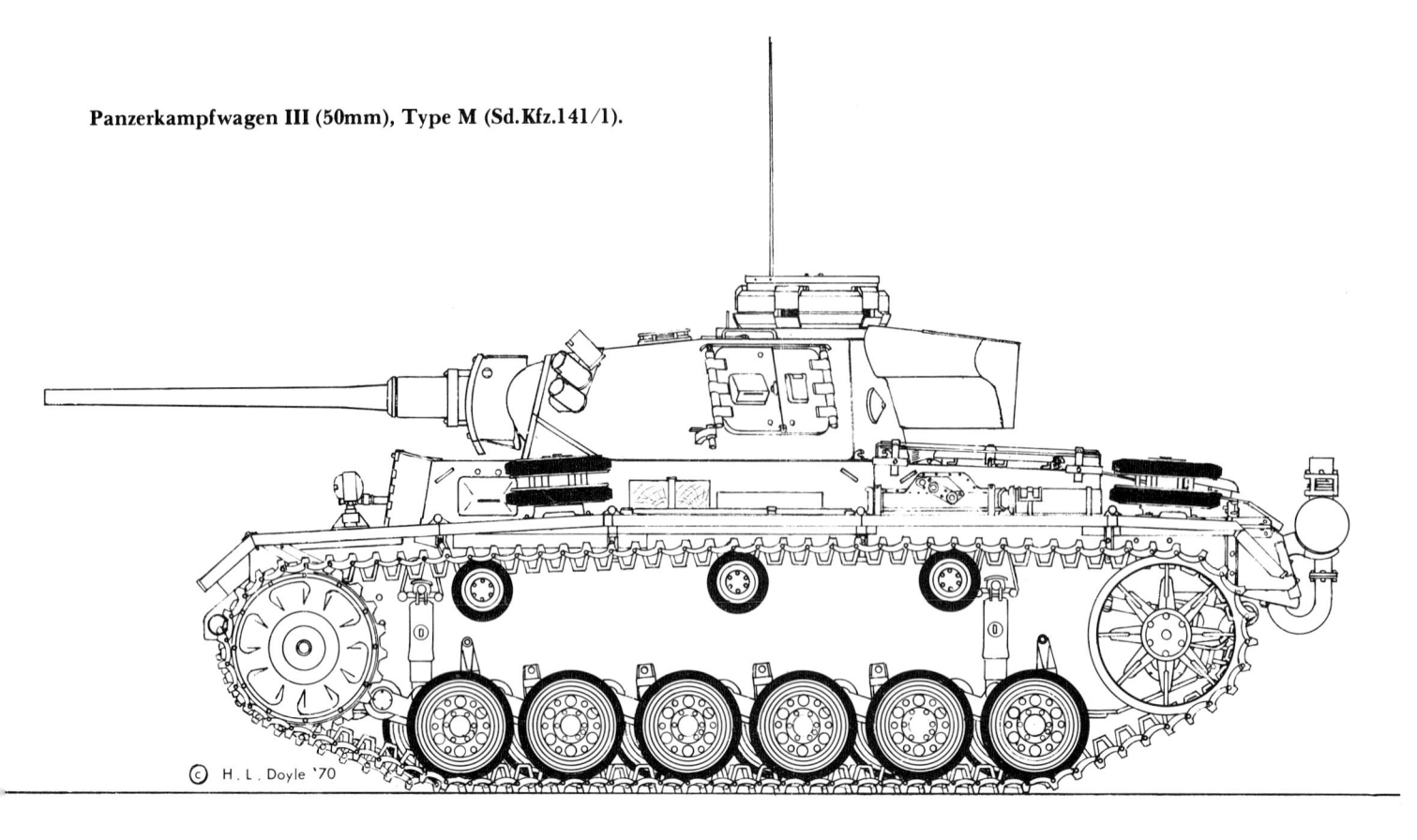

© H.L. Doyle '70

**This drawing shows the structure of the new exhaust system.**

Tanks of the last version were designated Panzerkampfwagen III, (75mm) Ausf. N (Sd.Kfz.141/2). The Daimler-Benz designation was "11/ZW", and chassis numbers began with 78 001. These tanks were fitted with the 75mm KwK L/24 gun and carried 64 shells for the tank gun plus 3450 rounds of machine-gun ammunition. The fighting weight was 21.3 tons. Minor changes to the ammunition holders made possible the retention of the 50mm facilities. The divided hatch cover atop the commander's cupola was replaced by a one-piece round cover. The servicing ports in the upper bow armor now had to be unscrewed from inside, as the hinges had been eliminated. According to Army Technical V-Blatt 1943, no. 789, the spotlights on all armored vehicles had to be attached so as to be removable. From Ausf. M on they had already been mounted on both fenders, while in all other versions they had been attached to the upper bow plate.

In 1943 there were still 349 Panzer III tanks produced. In August of that year, production of the Panzer III was officially halted, and the facilities thus freed were con-

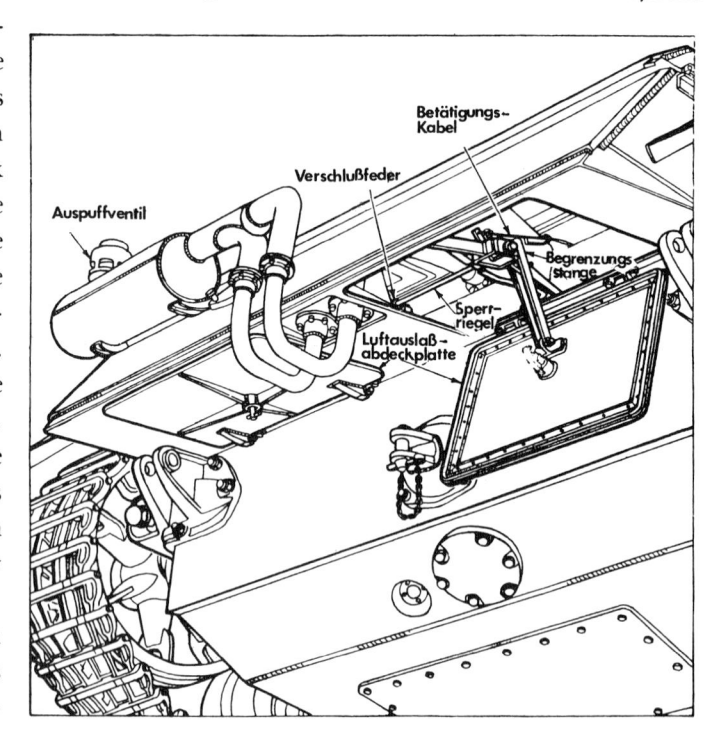

The most vulnerable parts of any tank are its tracks. Changing track links was practiced frequently to assure the readiness of the vehicles. This picture shows the use of special tools to reconnect the track.

verted to the production of Sturmgeschütz.

At Rügenwalde on March 19, 1943 a Panzer III tank with side aprons was first displayed. The aprons served to strengthen the side armor protection, especially against antitank and HL weapons. These side plates, 5 mm thick, could be removed from longitudinal bars on both sides of the tank and attached rigidly around the turret (according to Army Technical V-Blatt 1943 no. 433). The attachment of these aprons increased the width of the vehicle to 3410 mm.

A protective coating of Zimmerit, the use of which was ordered by Army Technical V-Blatt 1944 no. 733, was meant to prevent the attachment of magnetic charges. This paste-like substance was applied with a trowel and hardened with a blow-lamp. After the end of 1944 this substance was no longer used.

At the beginning of development, German armored vehicles were equipped with a smoke-laying device, which was screwed onto the closing panel of the rear armor. The five smoke cartridges hung on chains could be

**Panzerkampfwagen III (Fl), Ausf. M (Sd.Kfz.141/3).**

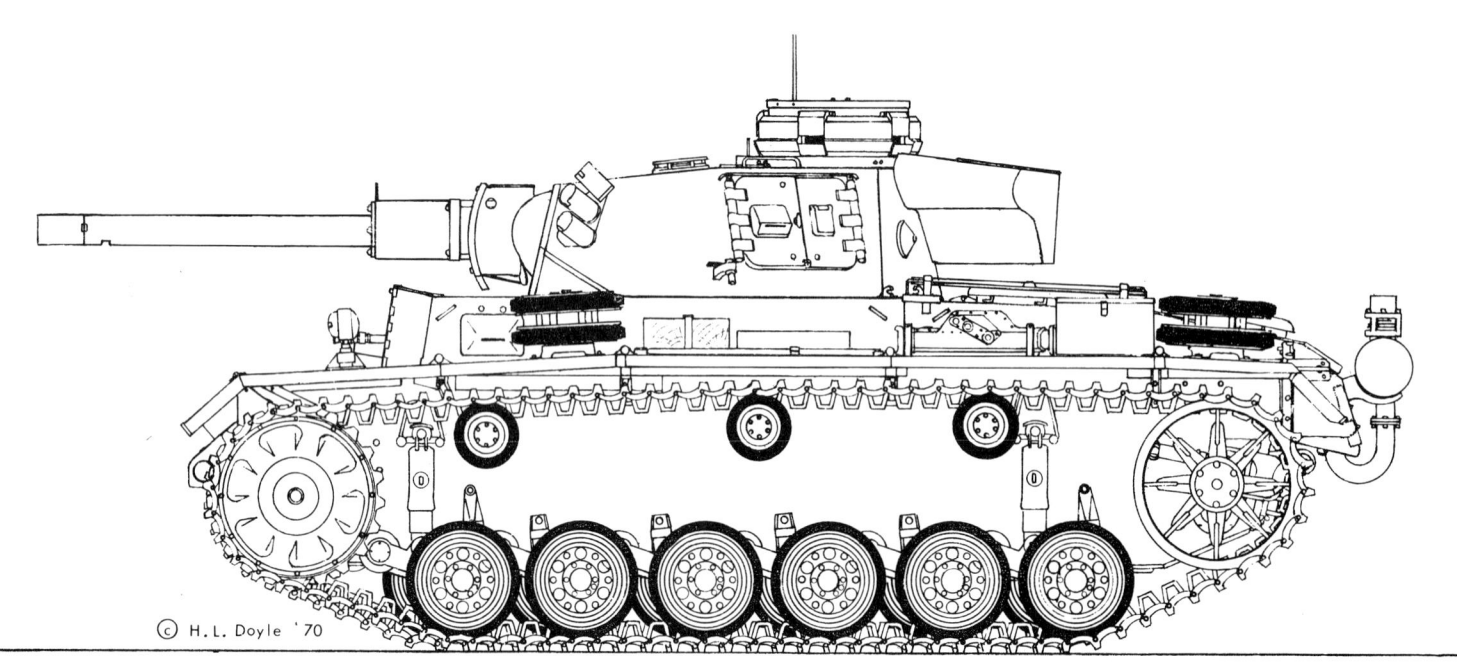

© H.L. Doyle '70

As of 1942, large numbers of the 75mm KwK L/24 were used as the primary armament of the Panzer III/ When tanks of older types were repaired, they were rearmed with the new gun. This picture shows a Ausf. L vehicle with the new primary armament.

100 Ausf. M vehicles were converted to flamethrowing tanks by the Waggonfabrik Wegmann.

These vehicles were basically the same as the battle tanks. Instead of the tank gun, a flamethrower was installed.

Because of the insufficient range of these flamethrowers, the vehicles were quickly withdrawn, and some were refitted with tank guns.

The final version of the Panzerkampfwagen III was equipped with the 75mm KwK L/24. It also differed in having a one-piece hatch on the commander's cupola. As of Ausf. M, the Panzer III was fitted with removable headlights attached to the track covers. This picture shows the surrender of German armored vehicles to the British Army in Norway after the war ended.

**Panzerkampfwagen III (75mm), Ausf. N (Sd.Kfz.141/2).**

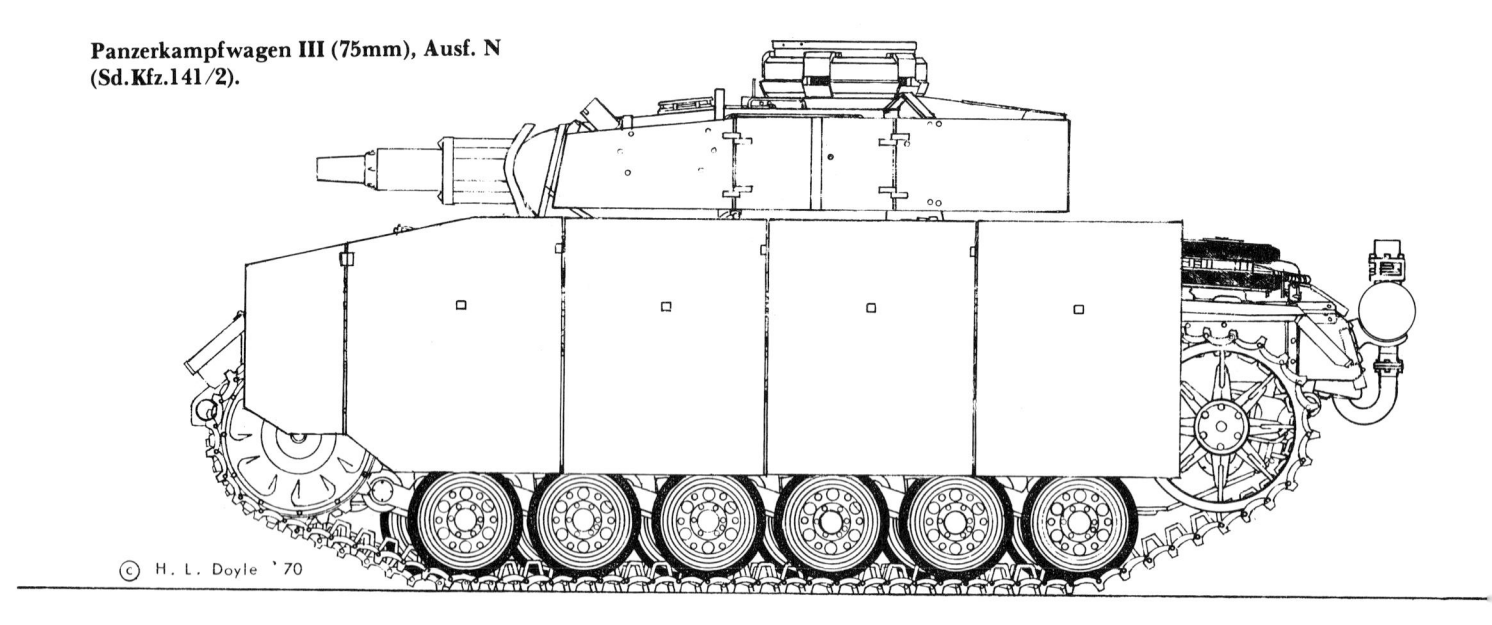

fired from inside the vehicle. As of 1943, all new-type armored vehicles and all those brought back to Germany for repairs were equipped with smoke-laying launchers, which now consisted of three launchers, combined as one unit, on each side of the turret. The different location of the launchers produced a fanlike spread of the smoke. For the troops in action on the eastern front, a widened track, the so-called "Ostkette", was put into use as of 1944 (Army Technical V-Blatt 1944 no. 256). It was meant to give the Panzer III and its derivatives greater off-road mobility in snow and mud. This was a makeshift arrangement, as these tracks, because of their one-sided widening, could be used safely only in flat country. The loading width of the tank with Ostkette was 3266 mm.

An improved Fliegerbeschussgerät 42 (Anti-Aircraft Device 42) replaced the Fliegerbeschussgerät 41 in 1943. It could be attached to the commander's cupola with set screws (used for MG 34 and 42).

Like the Panzerbefehlswagen III, the Panzerbeobachtungswagen III (Sd.Kfz.143) was a variant of the Panzerkampfwagen III with special equipment for artillery observation and long-range communication vehicle. But the Panzerbeobachtungswagen III was only a stopgap measure used until the introduction of the VK 903, which was planned as the standard observation vehicle. In terms of long-range communication, the Panzer III vehicles were already equipped like the final vehicle and thus went into service in 1941 and 1942. Installed in them were:

a. the 30-watt medium-wave transmitter and receiver (Fu 8),

b. the medium-wave receiver (Fu 4),

c. the Funksprechgerät f,

d. the on-board speaker system, and

e. the backpack radio g (for use outside the vehicle).

Passing on the command to fire within the artillery units from the battery officer's vehicle to the self-propelled armored howitzers was done by using the Funksprechgerät f, for which a loudspeaker was attached to every armored howitzer on the battlefield, so that the entire gun crew could hear it. This eliminated the problematic wearing of headsets. Since the effective range of the Funksprechgerät f under very favorable conditions amounted to 4 or 5 kilometers, it was sometimes possible to carry on a battery's entire radio communication with this device. The entire firing position staff heard the information provided by the observer, while the conversion from meters to lines (range or elevation) was made

On March 19, 1943, Panzer III tanks with side aprons were first displayed. They served to strengthen the side armor protection against antitank weapons.

As of 1944, the Panzer III in action on the eastern front were fitted with wider tracks, the so-called "Ostkette." Because of the extra width, the tanks could be driven safely only on flat terrain.

Daimler-Benz built the so-called Panzerbefehlswagen from 1938 on, and it reached the troops in three versions. The vehicles differed externally in terms of the frame antenna over the engine. The vehicle was designated "**Panzerbefehlswagen III Ausf. D 1, (Sd.Kfz.266/267/268).**" They varied only in their radio equipment.

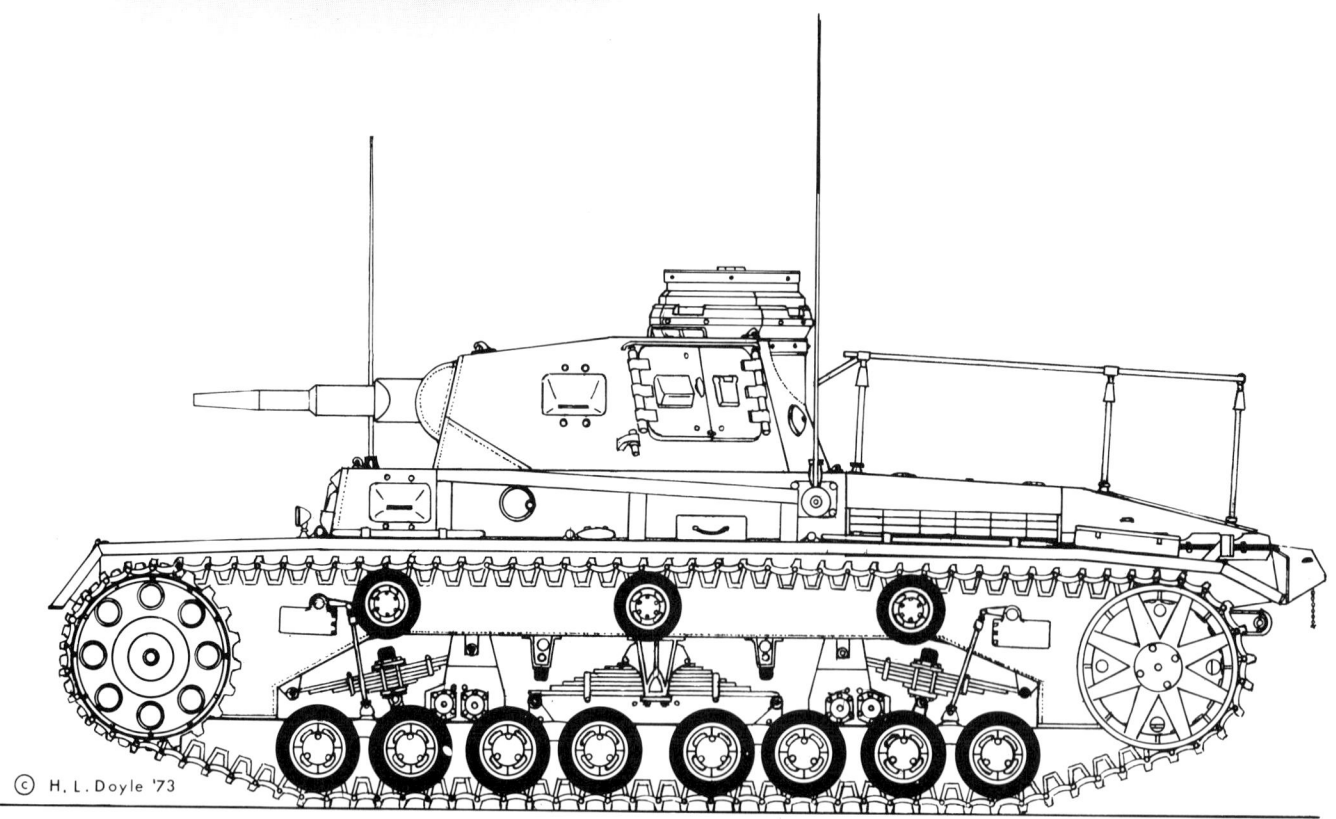

The Ausf. D 1 of the Panzerbefehlswagen III was followed by Types E and H, based on the same versions of the Panzerkampfwagen. The modified commander's cupolas of these two vehicles are easy to see.

Panzerbefehlswagen III, Ausf. E (Sd.Kfz.266/267/268).

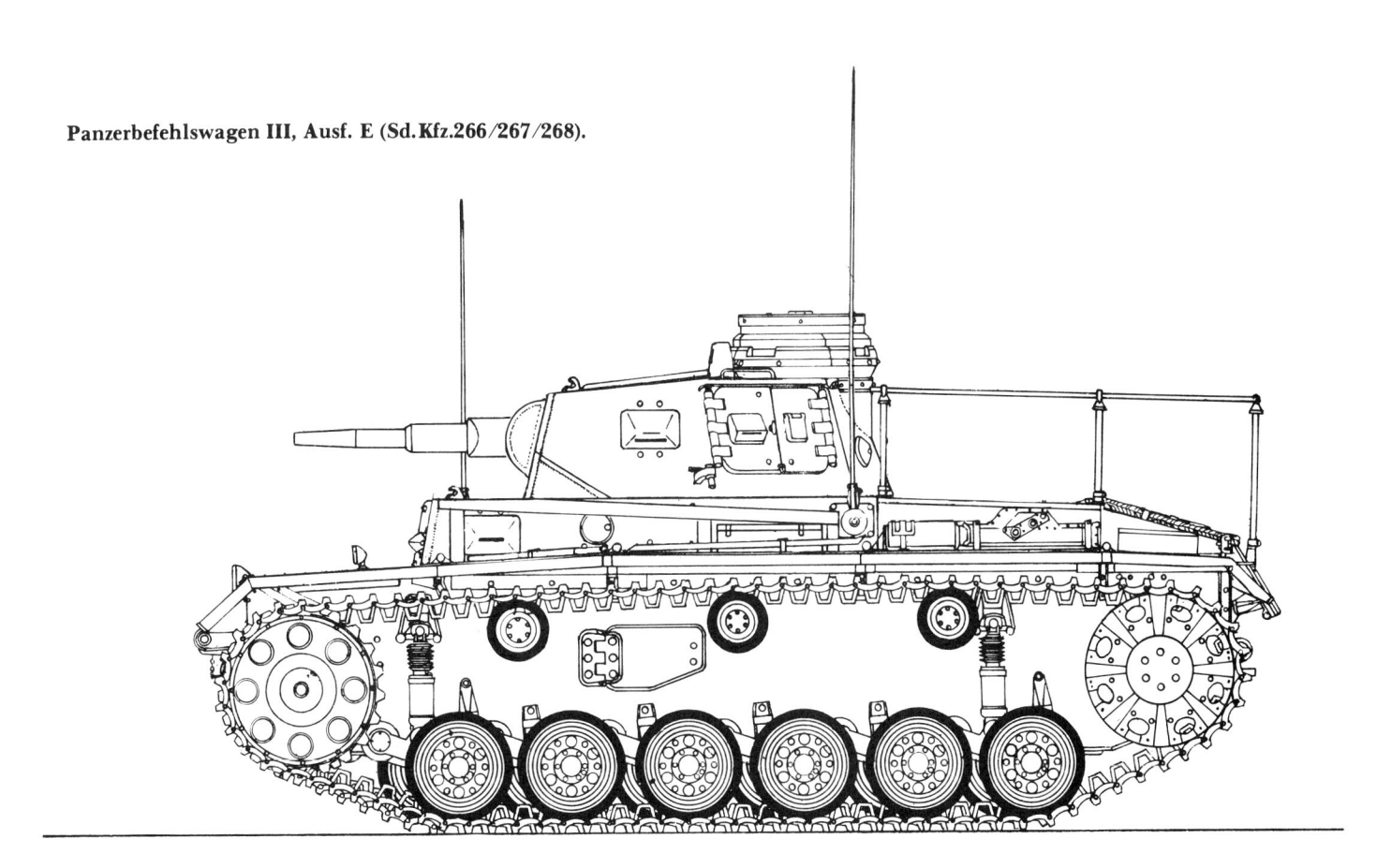

**Panzerbefehlswagen III, Ausf. H (Sd.Kfz.266/267/268).**

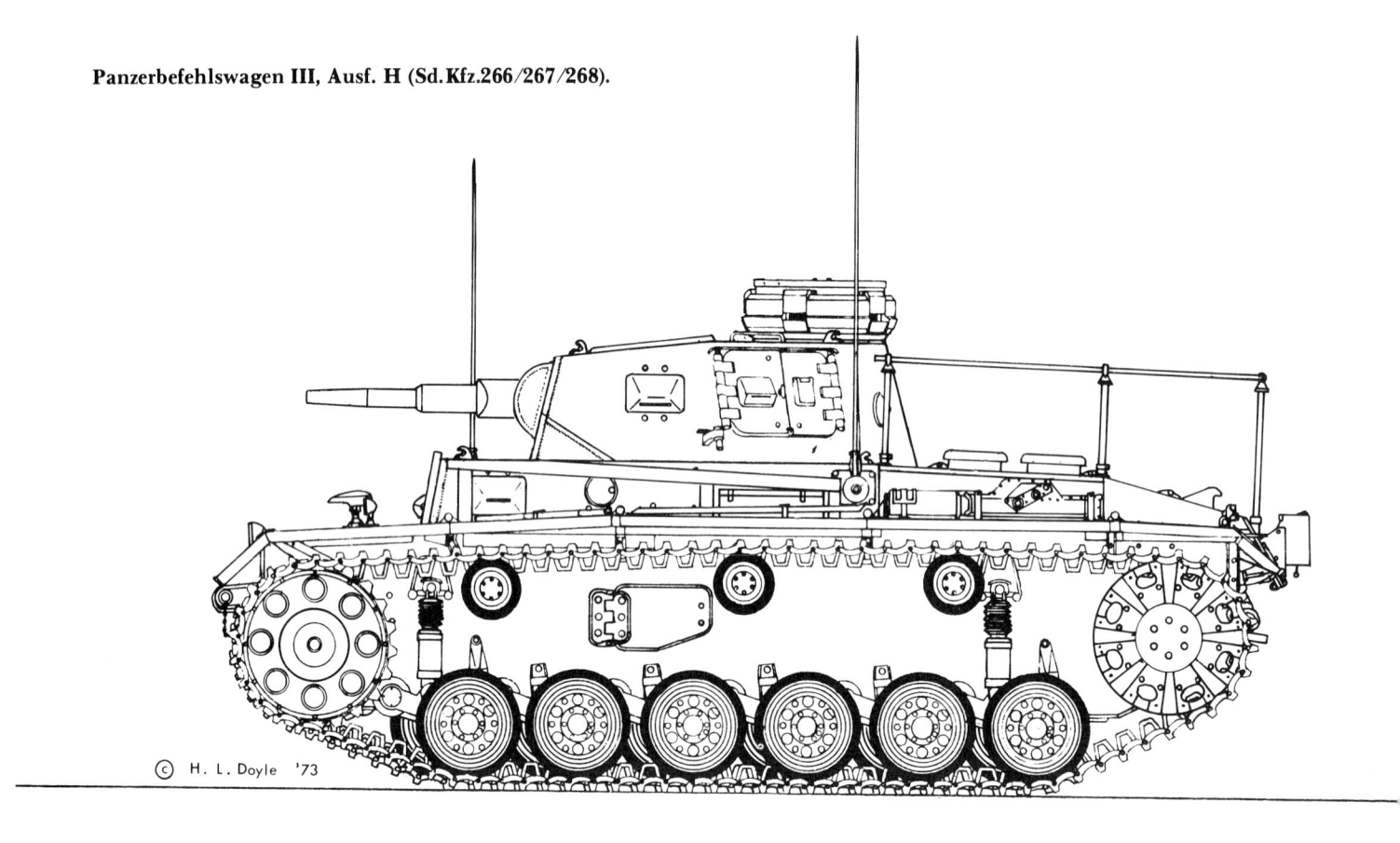

by the gun leader. This situation occurred rarely, though, as the observers needed the Funksprechgerät f for their communication with the Panzer units or Panzer grenadier battalions. The 30-watt medium-wave devices (Fu 8) were highly valued by the troops because they provided reliable communication to a distance of 20 kilometers and rarely malfunctioned. Their main disadvantage was that bearings on them could be taken more easily than on the UKW devices, though this was no great problem because of the Panzer artillery's mobile style of combat, and the troops accepted it in view of the devices' reliability.

The Panzerbeobachtungswagen III had a five-man crew (observer, assistant observer, driver and two radio-operators) and was equipped with an observation periscope that was activated from the assistant observer's seat. In the rotating turret was an MG 34 in a ball mantelet. Instead of a tank gun, the vehicle carried a similarly shaped sheet-metal dummy barrel. A bow machine gun was not included, but the radio-operator in the bow could fire his machine pistol through a port.

In 1943 several vehicles of an improved Beobachtungswagen III version went into service. These vehicles were fully armed with the 50mm KwK L/60 but had, in addition to the already mentioned radio equipment, additional viewing ports on the sidewalls of the upper hull armor as well as the turret.

The armored howitzer batteries were supplied, as of 1943, with two observation tanks and one fire-control vehicle, which used the same radio equipment. This enabled them to exchange duties if necessary. This radio equipment already corresponded to the "final solution" striven for by the Panzer artillery during the war. It proved itself fully from 1943 to 1945.

A Command Vehicle III of the 13th Panzer Division in Russia. The picture shows the added antennas. The turret is fixed in position, and the primary weapon is a dummy.

This picture shows a Command Vehicle III with winter paint leading a column of Panzer III tanks.

For armored vehicles with a gross weight of up to 22 tons (Panzer III and IV with their variants), the firms of Busch in Bautzen, Ackermann in Wuppertal, and Fink in Stuttgart built a 22-ton low-loader trailer (Sd.Anh.116) until 1945. The trailer weighed 13.8 tons, and the allowable total weight was 36.8 tons. The tire size was 13.50-20. The price was 28,000 Reichsmark per trailer. The trailer was developed by the Lindner firm of Ammendorf, near Halle, which originally developed the experimental 642 trailer. The tractor normally used was the 18-ton half-track tractor. There were 141 of these trailers in use by the army on April 1, 1942.

On May 25, 1938, the Daimler-Benz AG received a contract to develop a new tank of the 20-ton class, which was later supposed to replace the Panzer III. The new vehicle bore the Daimler-Benz designation of "ZW 40" and was regarded as the immediate successor to the "ZW 38." Prototypes of this 22-ton vehicle were built and tested thoroughly in 1939 and 1940. Our commentary on the Panzer III cannot be ended without a reference to the Panzerkampfwagen VK 2001 (DB). The powerplant was a Daimler-Benz 12-cylinder Diesel engine of the "MB 809" type, and its power was transmitted to the front drive wheels by a DB-Wilson 8-speed pre-selector gearbox. A ZF 8-speed pre-selector gearbox was also available optionally. The first prototype had a 3- or 4-radius overlapping drive,

**The final version of the command vehicle based on the Panzer III was Ausf. K. The vehicle carried a 50mm KwK 39. Instead of the frame antenna, a "star" antenna, harder to recognize, was now fitted.**

A Panzerbeobachtungswagen III in winter camouflage paint.

In the mantelet there was a ball mount for an MG 34 for close combat use.

In Panzer Artillery units the Panzerbeobachtungswagen III (Sd.Kfz.143) was used. This vehicle had upgraded radio equipment; its gun was only a dummy.

75

**Panzerbeobachtungswagen III (Sd.Kfz.143).**

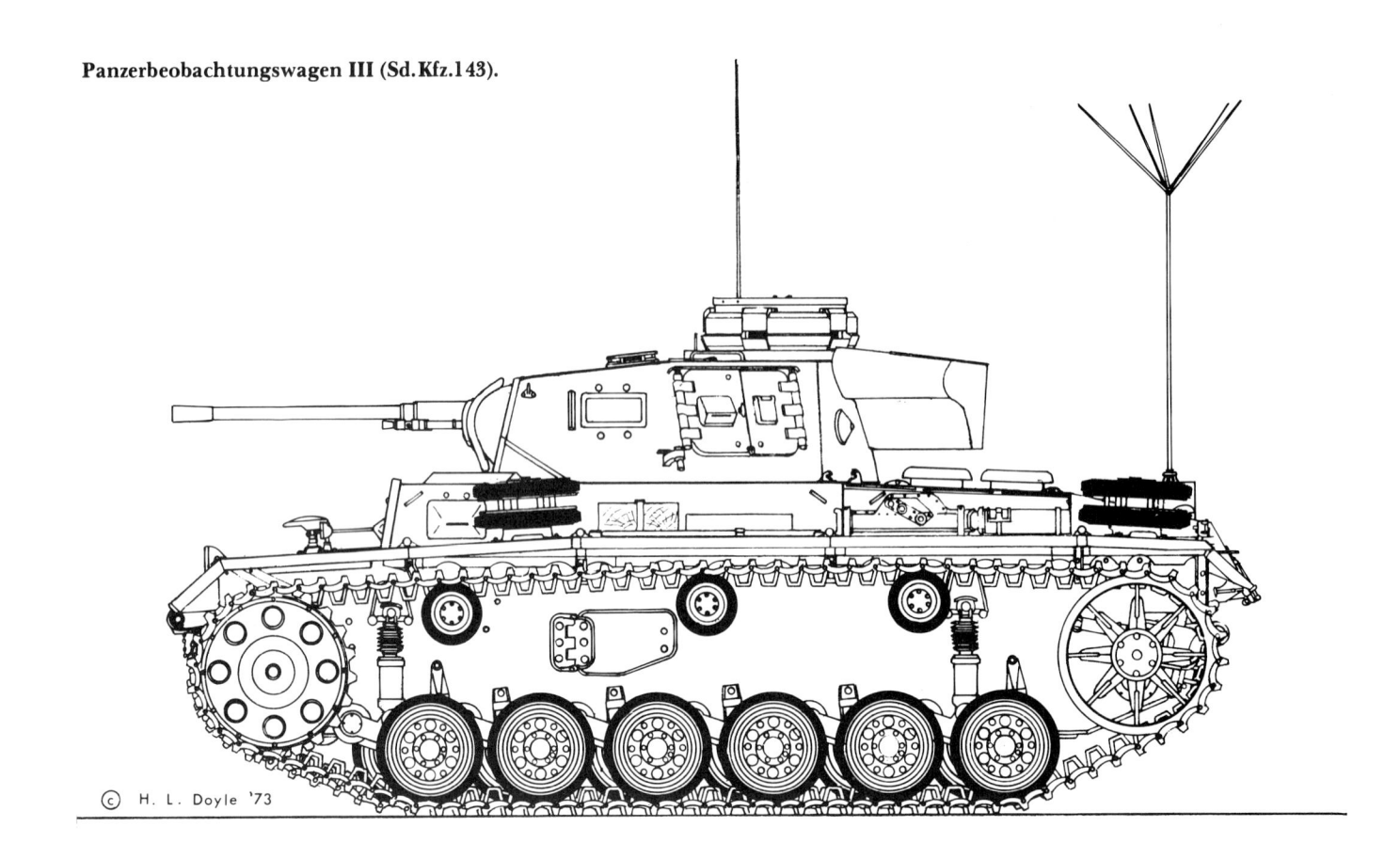

© H. L. Doyle '73

These vehicles were fitted later with the obligatory armored aprons. The additional antennas are easy to see.

The 22-ton Low Loader Trailer (Sd.Anh.116) was available to transport the Panzer III and IV tanks and their variants. The towing vehicle was usually the 18-ton Towing Truck (Sd.Kfz.9).

Special Trailer 116, Type Ba 38, was manufactured by the firms of Busch in Bautzen, Ackermann in Wuppertal, and Fink in Stuttgart.

the second a Daimler-Benz clutch steering drive. The 7-wheel running gear was suspended on leaf springs. The top speed was 50 kph. The vehicle's dimensions were: length 5130 mm, width 3020 mm, height (without turret) 1640 mm. The length of the tracks was 2757 mm, their width 440 mm.

A discussion at Works 40 in Marienfelde during 1938 centered on the question of a motor for the coming tank. For the "VK 2001 (DB)" tank described above, a 400-horsepower motor was needed, and the WaPrüf 6 (Army Weapons Office) immediately suggested a Maybach Otto-type motor. Nothing confirms the monopolistic position that Maybach had attained in German tank motor construction better than the fact that this firm delivered some 140,000 motors, with a total power of 40 million horsepower, to the Wehrmacht during the war. The Daimler-Benz AG constantly suggested Diesel engines of their own manufacture to the Army Weapons Office, and for the "VK 2001" (DB) they decided to develop a new Diesel engine. This powerplant, built at Works 60 since June of 1938, was designated "MB 809", was a V-12 engine of 25.5-liter displacement, produced 400 HP at 2100 rpm, and took up no more space than the Maybach HL 190 motor (12 cylinders, 10-liter displacement, 2400 rpm, weight 1000 kilograms) proposed by the Army Weapons Office. Constant changes to the tank that was supposed to go into series production brought with them corresponding adaptations of the Diesel engine's design.

The Daimler-Benz AG tried in numerous ways to save space and thus weight in the motor to apply to the armor. But neither the transverse mounting of the engine nor the use of considerably more expensive welded steel cylinders provided advantages; the latter was intended to decrease the engine weight and make smaller water and oil capacities possible. The design work on "MB 809" was finished early in June of 1940, and the test run of the first motor took place in February of 1941, with the acceptance tests on March 12, 1941. This motor was already at Marienfelde on March 21, and was installed in the test vehicle. The test runs took place on the factory grounds there and in Kummersdorf. When it was learned after the beginning of the Russian campaign that the Soviets had come out with considerably heavier vehicles powered by much stronger engines, efforts were made to develop vehicles and engines with higher performance. The VK 2001 and the MB 809 Diesel engine were therefore not developed any further.

It should not go unmentioned, though, that the Army Weapons Office had occupied itself even earlier with plans for Diesel engines. The Austrian Saurer Works AG in Vienna had already received a contract in November of 1938 to build a 350-horsepower Diesel tank engine. The work on it began in August of 1940, and two such powerplants were actually produced (12 cylinders, 120 x 160 mm, weight 1400 kp).

Simultaneously with the work on the "MB 809", the Klöckner-Humbold-Deutz AG of Cologne was working on the design of a Diesel engine for the "VK 2001." This powerplant, contracted for by WaPrüf 6 (1c), was supposed to be ready for delivery by the end of 1941. It was an eight-cylinder radial engine that produced 350 HP at 2500 rpm and weighed some 450 kp. A test model was under construction, but series production never took place, as the work on this engine, like that on the Saurer Diesel engine, was halted for the same reasons as that on the "MB 809" tank.

**The successor vehicle to the Type "ZW 38" was to be the VK 200i (DB) tank designed by the Daimler-Benz AG. Only prototypes were ever built.**

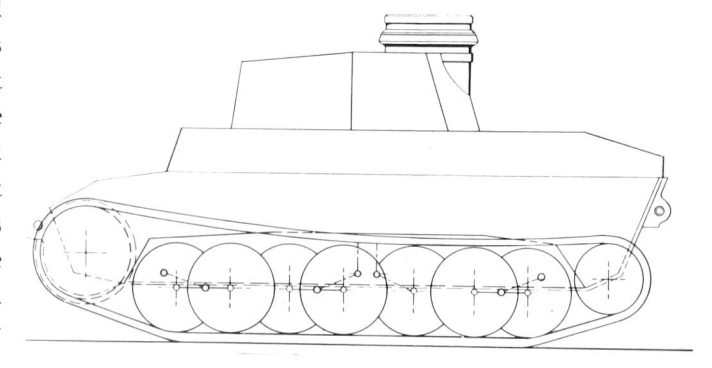

# Sturmgeschütz

The development of self-propelled gun carriages begun under contract from the Reichswehr Ministry in 1927 included, as a heavy version, the mounting of a 77mm infantry gun on a WD tracked chassis. The vehicle and gun were partially protected by armor. In the process, the concept of the "assault batteries" of World War I was taken up again. These attempts were halted temporarily in 1932, since other matters of military motorization seemed more urgent.

The later Feldmarschall von Manstein then suggested in a 1935 memo to the Chief of the General Staff and the Army High Command that the World War I concept of infantry escort batteries be taken up again. He envisioned a solution in the form of an armored self-propelled gun carriage for immediate support of the infantry. He suggested that every infantry division be assigned an assault artillery unit with three batteries of six guns each. His suggestion — not without initial resistance from numerous offices of the OKH — finally received the approval of the Chief of the General Staff and the Army Commander, Generaloberst von Fritsch.

The 8th (technical) Department of the General Staff, under the leadership of then-Oberst i.G. Model, was commissioned to work out appropriate developmental requirements for the Army Weapons Office and to carry the development forward. The First Referent of the 8th Unit, then-Hauptmann Hans Röttiger, was responsible for this assignment.

The official orders for "escort artillery under armor for infantry and antitank defense" came from the AHA (Insp. 4) in the form of Document 449/36 gKdos to the Army Weapons Office, dated June 15, 1936. The following developmental requirements were established:

The armament to be used should be a gun of at least 75mm caliber.

The range of traverse should be, if possible, more than 30 degrees without turning the vehicle.

A possible elevation of the barrel was to be provided for, in order to achieve a maximum range of 6000 meters.

The gun was required to have the armor-piercing capability to penetrate all armor thicknesses then known at a range of 500 meters.

All-around armor was required, with a superstructure open at the top and without a turret. The front armor was to be secure to 20mm with a 60-degree angle of impact.

The overall height of the vehicle should not exceed that of a man standing upright.

The other dimensions were to result from the size of the armored chassis utilized.

Other requirements were established as regarded the gun's supplies of ammunition and communication devices, as well as the number of crewmen.

Long, tedious negotiations were necessary to meet these requirements. Among other things, the open superstructure was finally closed, in order to protect the crew better in close combat. For the initial considerations as to the total concept, several Panzer II and IV chassis were taken from the production lines. With these makeshift measures, the first firing tests were carried out at the Army Weapons Office's shooting range in Kummersdorf in the winter of 1936-37. In addition to technical and tactical problems that surfaced during development, the basic question as to the appropriateness of this new weapon had to be decided on:

It made sense to turn this weapon over to the infantry, as these vehicles were to serve chiefly for direct support of the infantry divisions. But it was mainly a logistic consideration that made delivering these vehicles to the infantry impossible.

General Guderian declined to assign tanks for this purpose because he saw it as a weakening of his own plans for development, a view that was easy to understand given the poor production capability of the German tank industry.

The Artillery Inspection was still trying at this time to clear up the question of whether it might not be more purposeful to make the new escort guns horsedrawn as in World War I. In spite of this, there were those who suggested that the technical as well as the firing training involved would be left most safely in the hands of the artillery.

The General Staff decided in the end that the assault artillery units should be established as army artillery.

The first experimental assault guns used the eight-wheel running gear of the Type 2/ZW. Round servicing hatches were fitted in the bow armor.

Panzerkampfwagen III (37mm) (Sd.Kfz.141), Ausf. A

Panzerkampfwagen III (37mm) (Sd.Kfz.141), Ausf. E

81

**Panzerkampfwagen III (50mm) (Sd.Kfz.141/1), Ausf. M, side view**

Ausf. M as seen from above

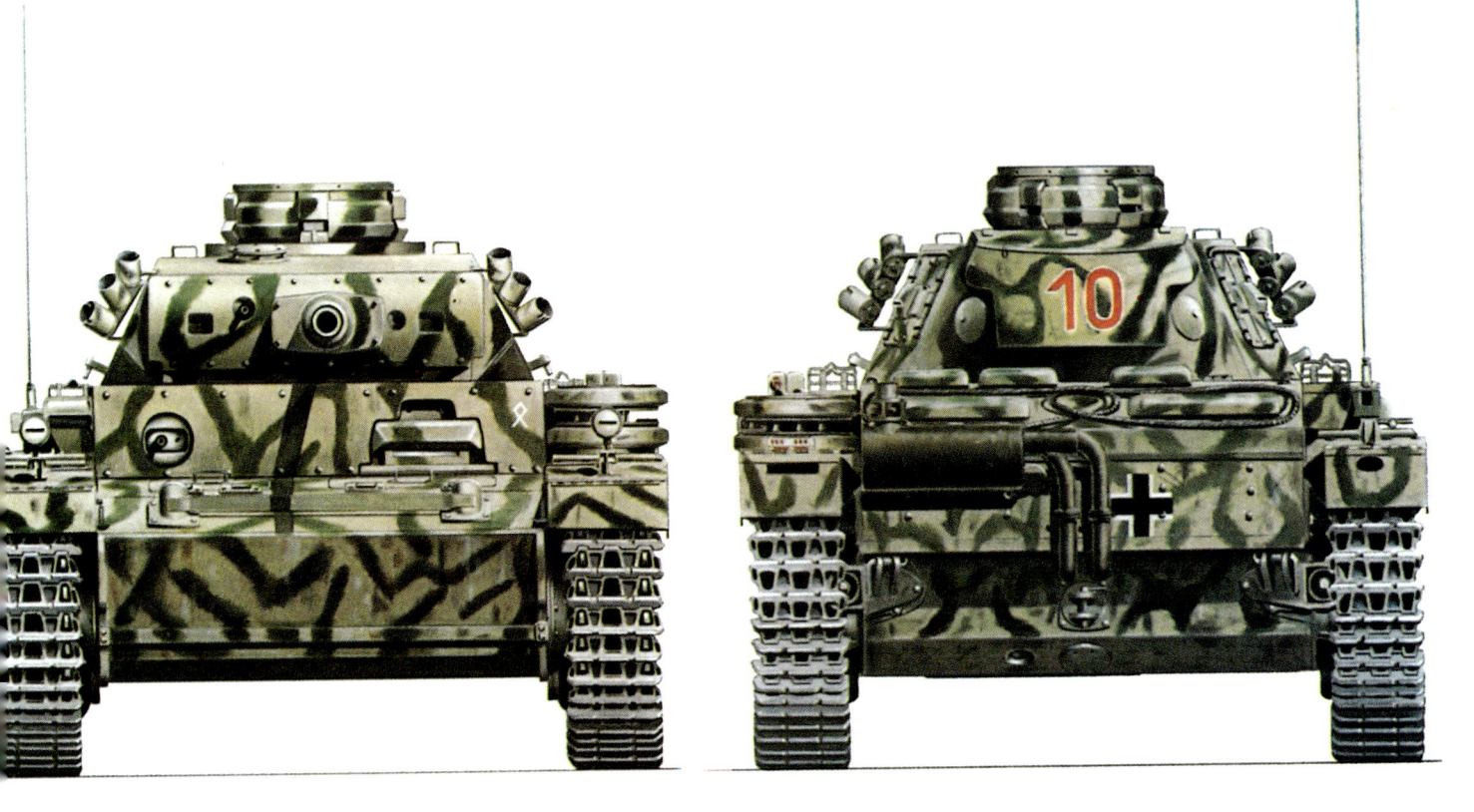

Ausf. M as seen from the front

Ausf. M as seen from the rear

Panzerkampfwagen III (75mm) (Sd.Kfz.141), Ausf. N

83

Gerpanzerte Selbstfahrlafette für Sturmgeschütz 75mm Kanone (Sd.Kfz.142), Ausf. D

Sturmgeschütz 40 (Sd.Kfz.142/1), Ausf. G

**Sturmgeschütz prototype on Panzer III chassis, Ausf. B (with soft steel superstructure).**

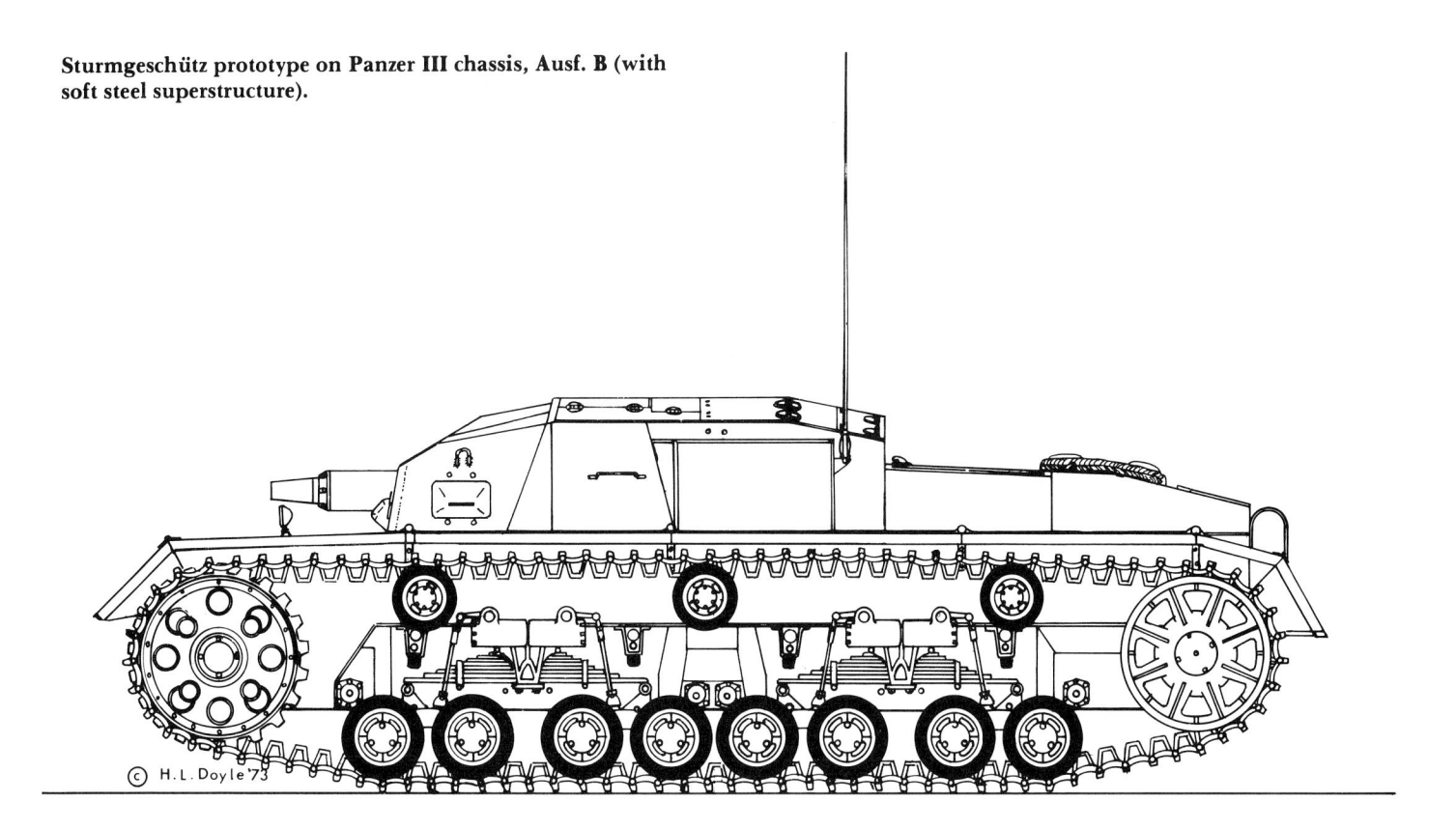

In 1936 the Artillery Training Regiment in Jüterbog had been given the assignment of working out the combat and utilization possibilities for an armored assault gun on a Panzer III chassis according to the guidelines of In. 4 (Artillery Inspection of the OKH). In 1937 a test battery, the 7./ALR, was established. In the spring of 1938 the first practical sessions in the country took place with the two available armored chassis and the dummy guns mounted on them. For reasons of secrecy, only a "37mm Pak (Sfl)" was mentioned at that time.

The Daimler-Benz AG in Berlin-Marienfelde was chosen as the developmental firm for the chassis and superstructure, while the Friedrich Krupp AG in Essen was made responsible for installing the guns. Through the use of an unchanged Panzer III chassis, there resulted a closed tank body armored with 50 mm plate in the front and 30 mm plate on the sides. The primary armament, a semi-automatic 75mm L/24 gun, was installed and had a traverse of 24 degrees and an elevation of -10 to +20 degrees. With a firing height of 1.5 meters, the entire vehicle was not to exceed a height of 1.95 meters. With a four-man crew and a supply of 44 shells plus an ultra-short-wave receiver, the gross weight amounted to 19.6 tons.

The development of these vehicles made good progress until the autumn of 1938. Five chassis were taken from the second experimental series, the "ZW" development (Type 2.ZW). These vehicles already had the finished appearance of the assault gun, but were recognizable by their 8-wheel running gear as well as the round hinged assembly openings attached to their bow armor. A viewing tunnel for the aiming optics was planned, while the later production version was fitted with the Rundblickfernrohr 32 (panoramic periscope 32) over the roof of the superstructure. The high armor plate that was added later on both sides of the gun-cradle armor was also lacking. A

**This version likewise had a sight tunnel with shot deflectors for the aiming optics.**

The light armored ammunition transport vehicle was introduced at the same time as the establishment of the first assault-gun units.

The lightly armored ammunition vehicle (Sd.Kfz.252), which was intended solely for use by the Sturmgeschütz units. The body was closed at the top.

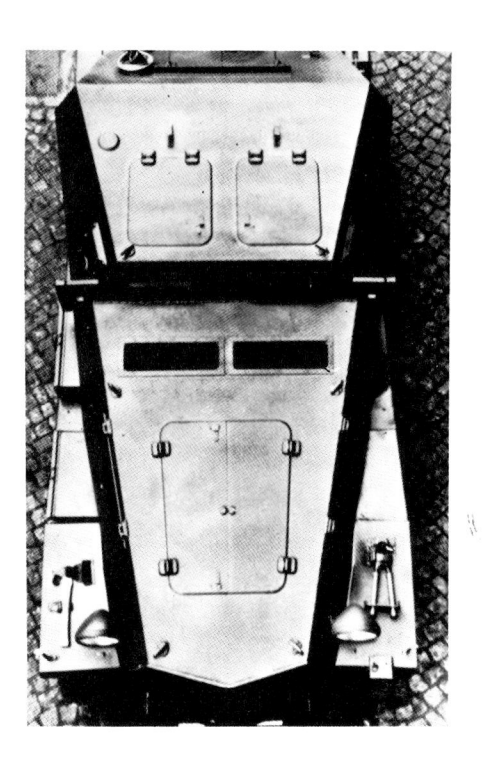

At the same time, the lightly armored observation vehicle (Sd.Kfz.253) made by Wegmann in Kassel was introduced as the command vehicle of the Sturmgeschütz units. This vehicle was also closed at the top.

87

driver's periscope was not yet included. For reasons as yet unknown, all training on assault guns was ordered halted in the autumn of 1938. Assault guns were not used in the Polish campaign of 1939.

Yet the experience gained in this campaign showed the need for such a weapon to support the infantry, which led to the emphatic request for "independent batteries" in the autumn of 1939.

The initial series of 30 units that was then begun at Daimler-Benz was already based on the "5/ZW" chassis, which was immediately followed by the first production version of the Gerpanzerten Selbstfahrlafette für Sturmgeschütz 75mm Kanone (armored self-propelled carriage for the 75mm assault gun). The chassis numbers for this Ausf. A (Sd.Kfz.142) started with 90 001. Their assembly was carried out exclusively by the "Altmärkisches Kettenwerk GmbH" (ALKETT) in Berlin. The firms that provided the armor plate during the course of production were: Brandenburgische Eisenwerke GmbH, Brandenburg/Havel, or Kirchmöser Works, Harkort-Eicken; Stahlwerke AG, Hannover-Linden; and Königs-und Bismarckhütte AG, Upper Silesia.

The vehicle itself consisted of the chassis, the armored superstructure attached to it, and the gun. The chassis and armored superstructure were separated by a dividing joint. The armored hull was designed to be the carrier of the chassis. The engine and transmission were mounted in it. In Ausf. A the drive train led from the motor through a tunnel in the fighting compartment to the gears in the bow area. The bevel gear with the steering gears were attached to the front of the gearbox. The armored hull consisted of the closed bow section, the open central section and the open rear section, made up of several pieces of armor plate of various thicknesses, welded together. In the bow were two boltable divided hatch covers, one of which served as an emergency exit for the driver. As in the Panzer III, the Maybach HL 120 TR engine and the Maybach Variorex pre-selector gearbox

**Beginning with the Ausf. A of the Sturmgeschütz, the 5/ZW chassis was utilized. This drawing shows the layout of the chassis.**

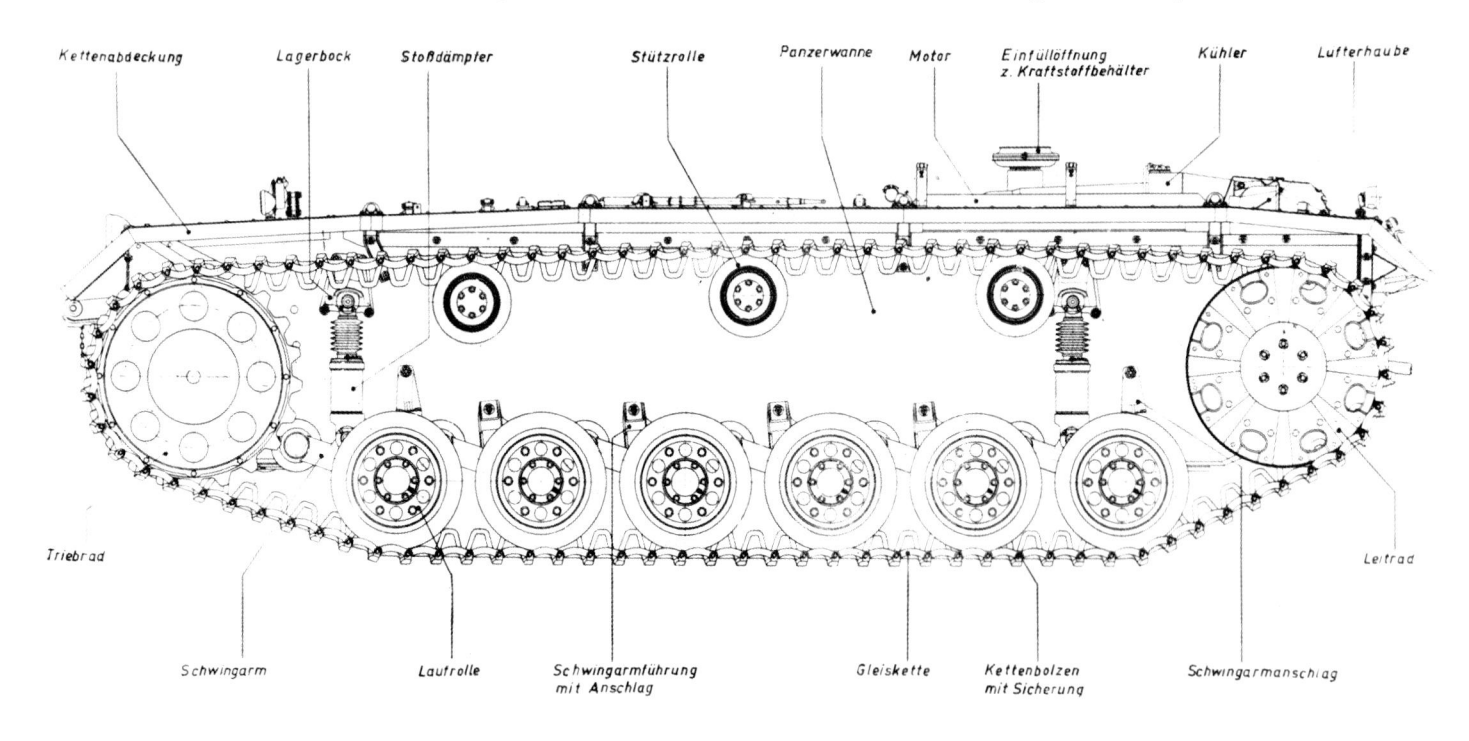

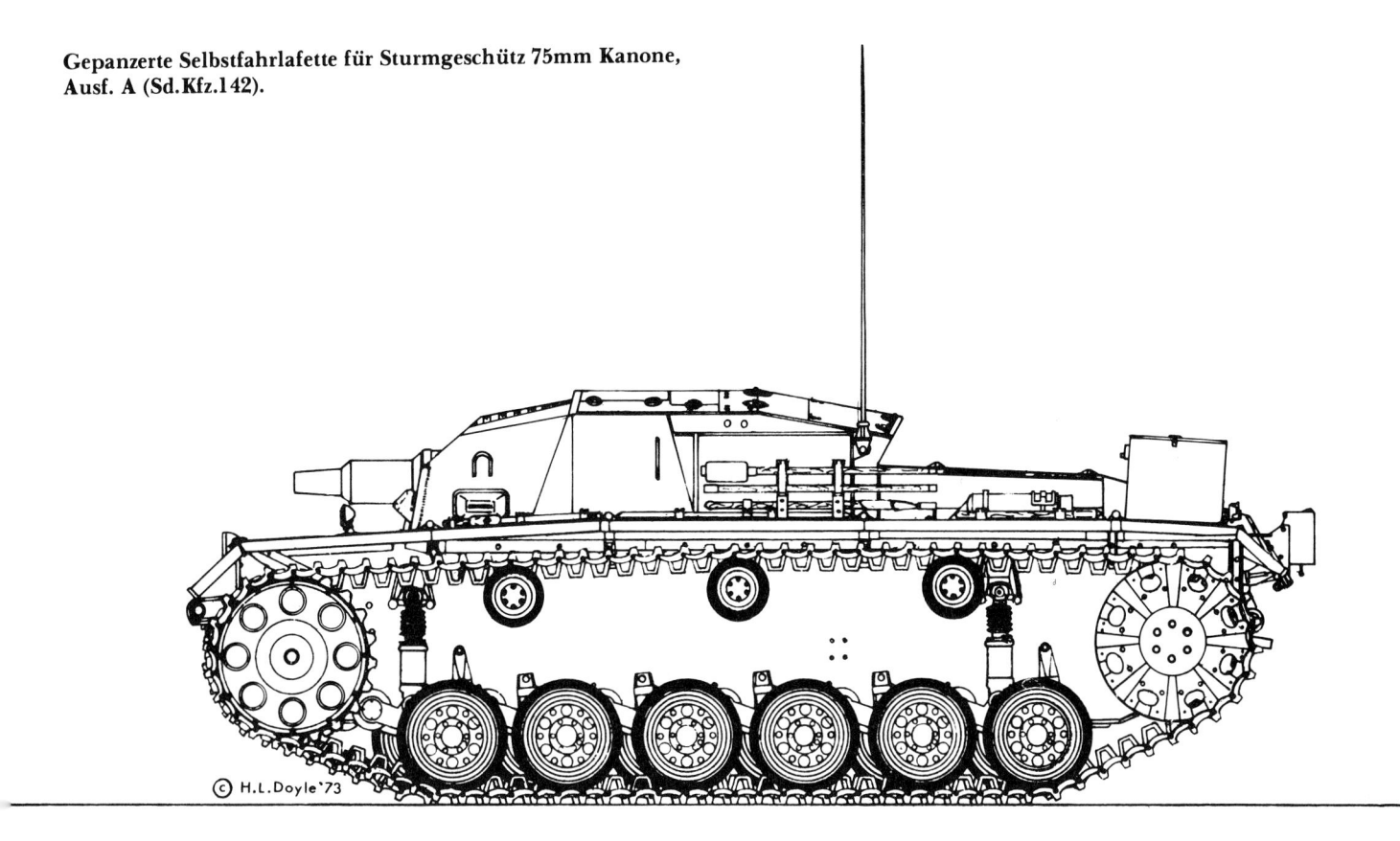

**Gepanzerte Selbstfahrlafette für Sturmgeschütz 75mm Kanone, Ausf. A (Sd.Kfz.142).**

were used. The gun rested on a pedestal. This consisted of two boxlike carriers that were linked with each other at the top and carried the baseplate that held the gun. The rear carrier was braced by two struts extending backward from the driveshaft tunnel. The floor, which was above the torsion bars, was made of textured sheet steel. The ammunition was housed in various boxes with angle-iron frames and sheet-metal coverings. In the empty space to the right of the gearbox was a container to hold two cases of three shells and two cases of two shells each.

The armament was the 75mm L/24 Sturmkanone, which had an actual barrel length of 1766.5 mm. The maximum shot range was 6000 meters. With the 75mm Gr. 38 HL/A shell, a maximum velocity of some 450 meters per second was attained. Explosive shells could also be fired. The aiming mechanism was a Rbfe 32 4x10-degree panoramic periscope with calibrations in meters. A view was gained through a cutout in the front armor of the fighting compartment. The equipment was completed by an ultra-short-wave radio receiver.

In the meantime, three further battalions (Battalions IV, V and VI) had been established in the Artillery Training Regiment. Of these, the VI./ALR, with three batteries (No. 16, 17 and 18), was made into an "Assault Gun Training and Replacement Battalion." By the beginning of the French campaign in 1940, six assault-gun batteries (No. 640, 659, 660, 665, 666 and 667) had been established, of which Batteries 640, 659, 660 and 665 took an active part in the campaign. Technical problems arose, particularly with the additional armor plate with which several of the assault guns had been equipped. The bolts needed to attach it loosened inside the vehicle under fire. Batteries 660, 666 and 667 were then prepared for Operation "Sealion" and carried out exhaustive loading drills in Holland.

In the meantime, on the basis of the good experience

**Vehicles of the first production series Ausf. A were already in use during the French campaign of 1940. They proved themselves well.**

with the assault guns in the French campaign, an official introductory contract had been placed in July of 1940. While the monthly production at this time amounted to some 20 units, as of September 1940 the production was to be increased to about 50 vehicles. In fact, monthly production as of November 1, 1940 had reached 30 vehicles, and in the whole year of 1940, 184 Sturmgeschütz had been built.

The chassis improvements to the Panzer III resulted in a change in the power transmission of the Ausf. B Sturmgeschütz, which appeared at the end of 1940, in that the driveshaft was now linked to the gearbox via a clutch. All other changes in the Type "7/ZW" were also adopted.

For the Ausf. C, which appeared in June of 1941, there were production difficulties. In all, 548 Sturmgeschütz were produced in 1941, and the monthly production figure was lowered to 40 in view of the increased monthly production quota of the Panzer III.

During the course of 1942, Ausf. D and E of the 75mm Sturmgeschütz were introduced.

In the optional use of the Ausf. E Sturmgeschütz as a platoon leader's gun, it became necessary to add an extension, like that on the left side of Ausf. A to D, on the right side as well to accommodate the radio equipment. Originally it had not been intended to provide the platoon leader with his own gun; like the battery chiefs and unit commanders, they were supposed to carry out their command duties in a lightly armored semi-track vehicle (Leichter Gepanzerter Beobachtung Kraftwagen Sd.Kfz.253).

The added extension on the side also provided the possibility of carrying six more rounds of ammunition. The supply thus added up to 50 rounds. A new change from the earlier types was an on-board communication system instead of a speaking tube. In addition, a rack for the MG 34 was provided in the right rear corner of the fighting compartment, as well as a rack on the rear ammunition chest (12 rounds) for seven drum magazines.

The platoon leader's Sturmgeschütz now had a 10-watt transmitter (10 WS h) and two ultra-short-wave receivers (UKW-Empf h) with a loudspeaker. This loudspeaker was intended for the aiming NCO, freeing him from carrying the double receivers. Later this loudspeaker apparatus was eliminated.

The chassis numbers for the Sturmgeschütz equipped with the short 75mm gun ran from 90 001 to 91 400.

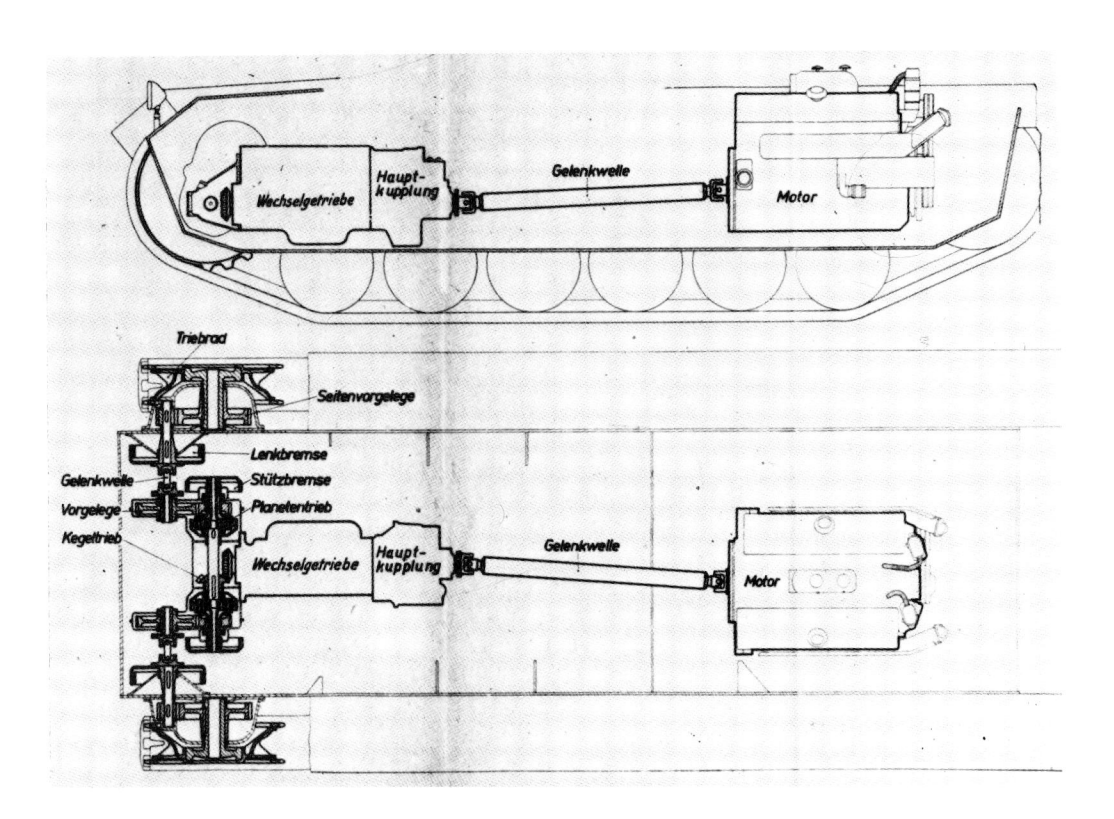

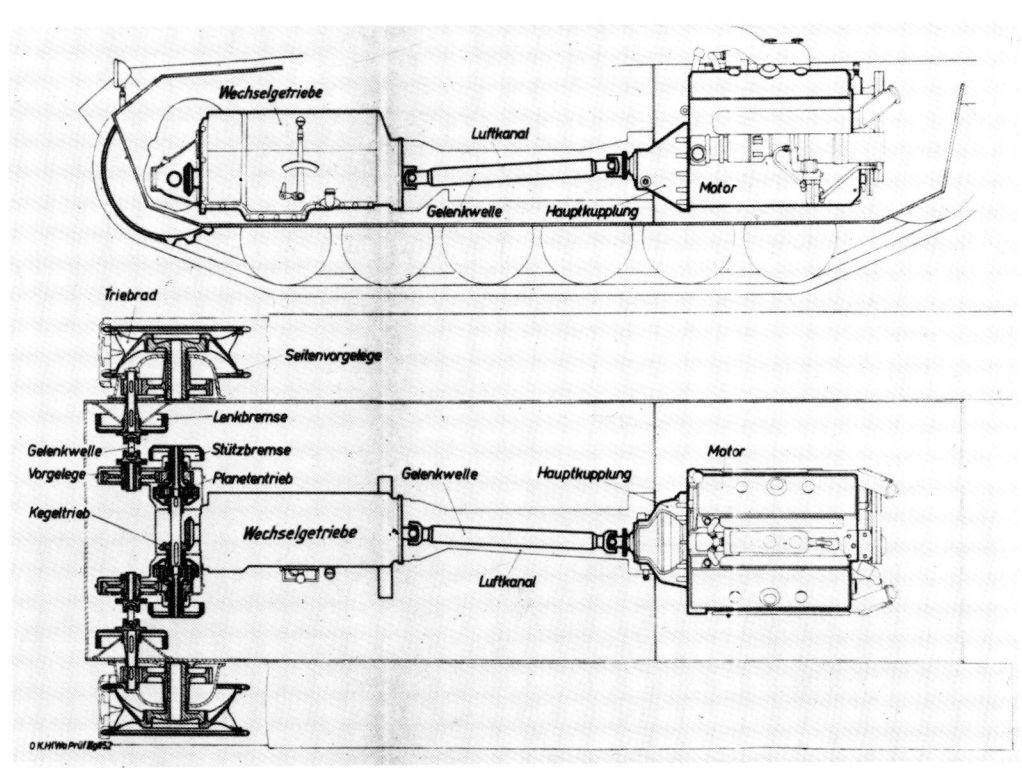

These drawings allow a comparison of the power trains of Ausf. A (above) and B to F of the Sturmgeschütz.

The round assembly ports in the bow plate had been eliminated from Ausf. A. The open sight tunnel was retained.

Gepanzerte Selbstfahrlafette für Sturmgeschütz 75mm Kanone, Ausf. B (Sd.Kfz.142).

The sight tunnel was eliminated, beginning with Ausf. B. The panoramic periscope 32 was located above the roof of the superstructure. Now the 7/ZW chassis type was utilized.

**Gepanzerte Selbstfahrlafette für Sturmgeschütz 75mm Kanone, Ausf C and D (Sd.Kfz.142).**

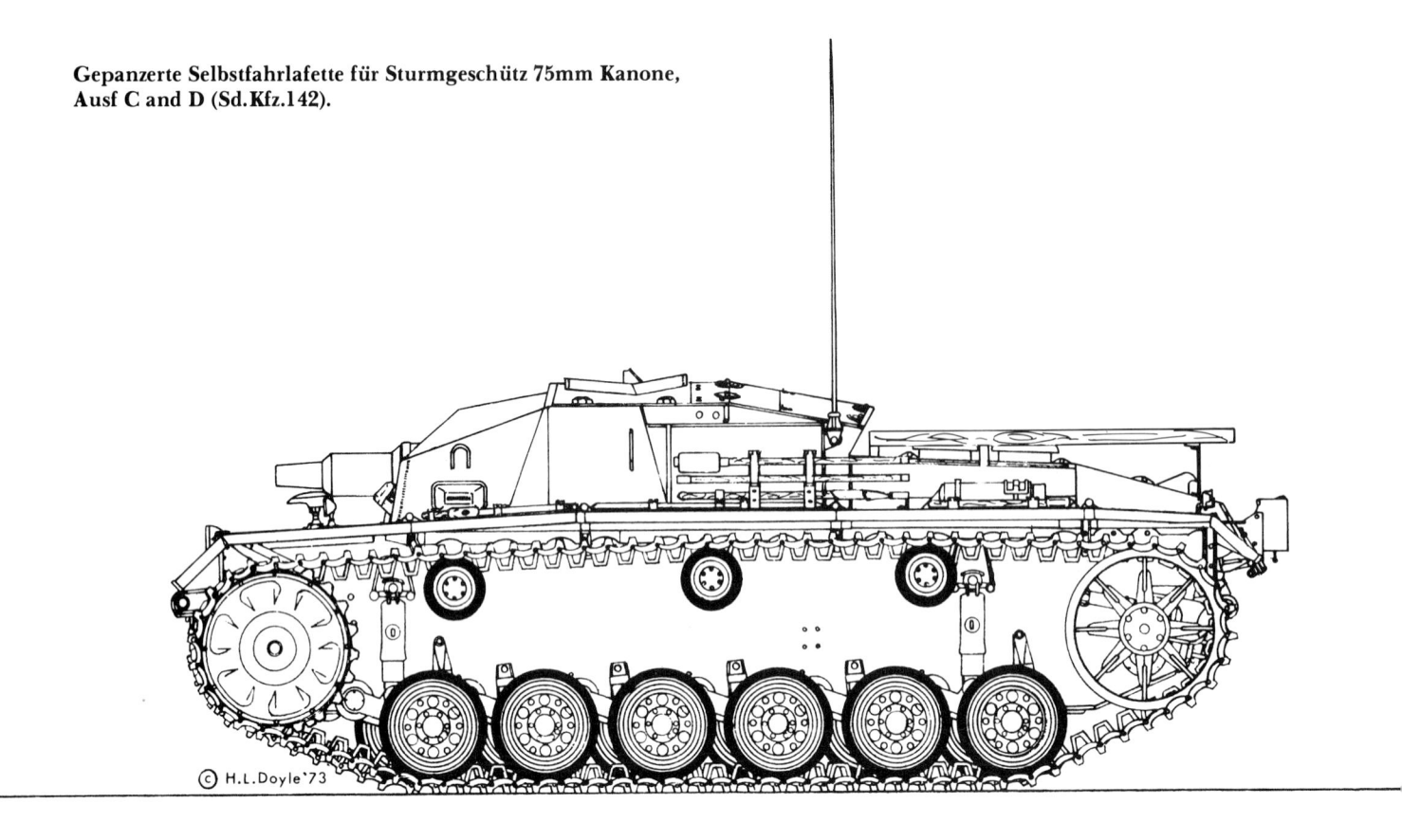

This side view of Sturmgeschütz shows the low superstructure of the vehicle as well as the short 75mm Sturmkanone. Spare torsion bars were sometimes stored on the sides of the hull above the road wheels.

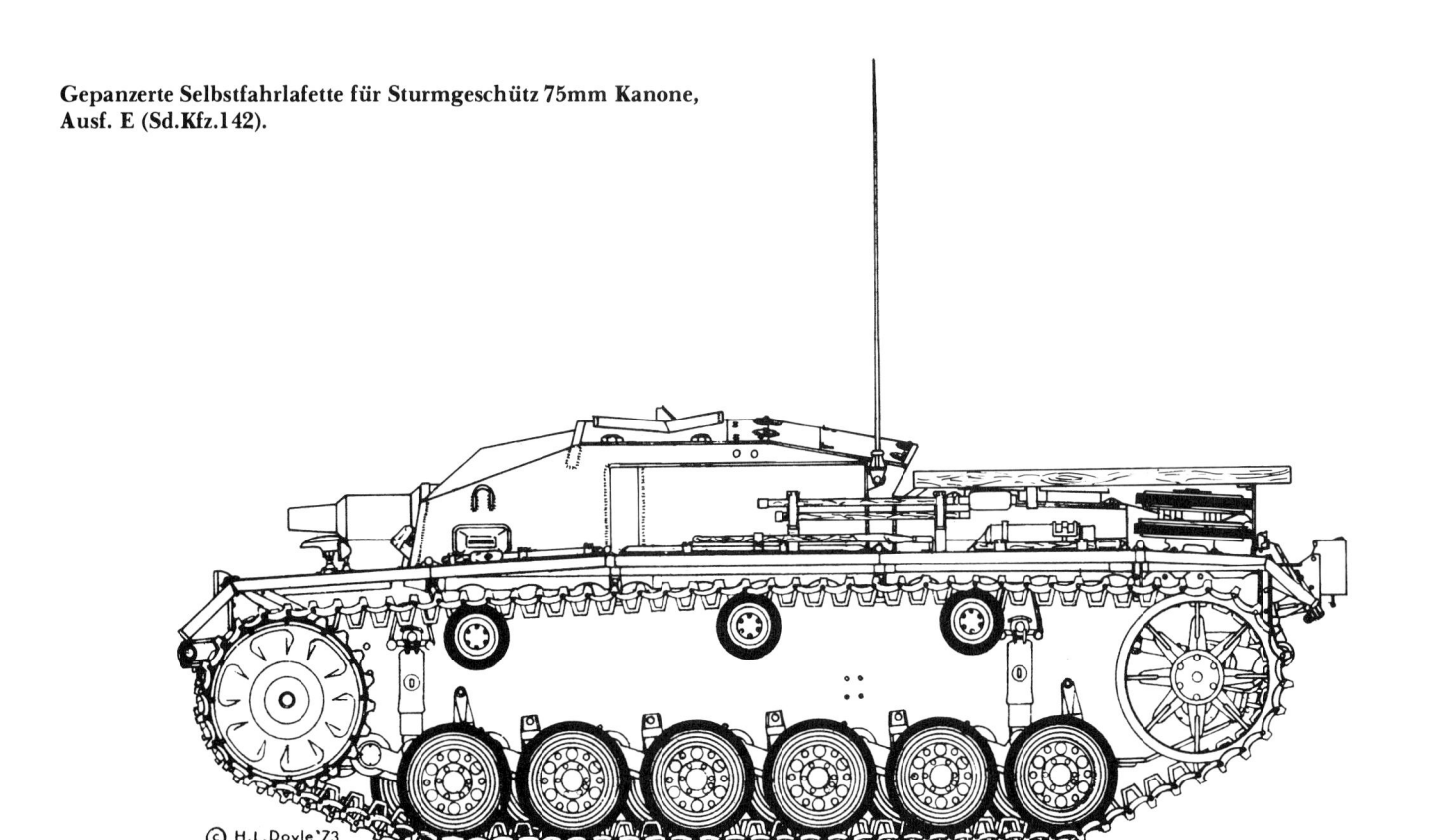

**Gepanzerte Selbstfahrlafette für Sturmgeschütz 75mm Kanone, Ausf. E (Sd.Kfz.142).**

Beginning with Ausf. E, the platoon leader's vehicles were fitted with an armored extension on the right side of the superstructure to house the additional radio apparatus. If the gun was used normally, there was thus room for six extra rounds of ammunition.

A Sturmgeschütz unit provides infantry support at the beginning of the Russian campaign in 1941. Armored halftrack vehicles were still used to supply ammunition.

The first Russian winter was a difficult test for men and machines. This picture shows an assault gun being towed out of the snow by a Panzer III.

The first three experimental Sturmgeschütz vehicles with a longer gun barrel appeared in February of 1942.

This weapon was the 75mm Sturmkanone 40 L/43.

This left-side view shows the otherwise unchanged superstructure of the Ausf. F Sturmgeschütz.

The same vehicle seen from the left rear shows the ventilator on the rear deck, which was standard equipment.

**75mm Sturmgeschütz 40, Ausf. F (Sd.Kfz.142).**

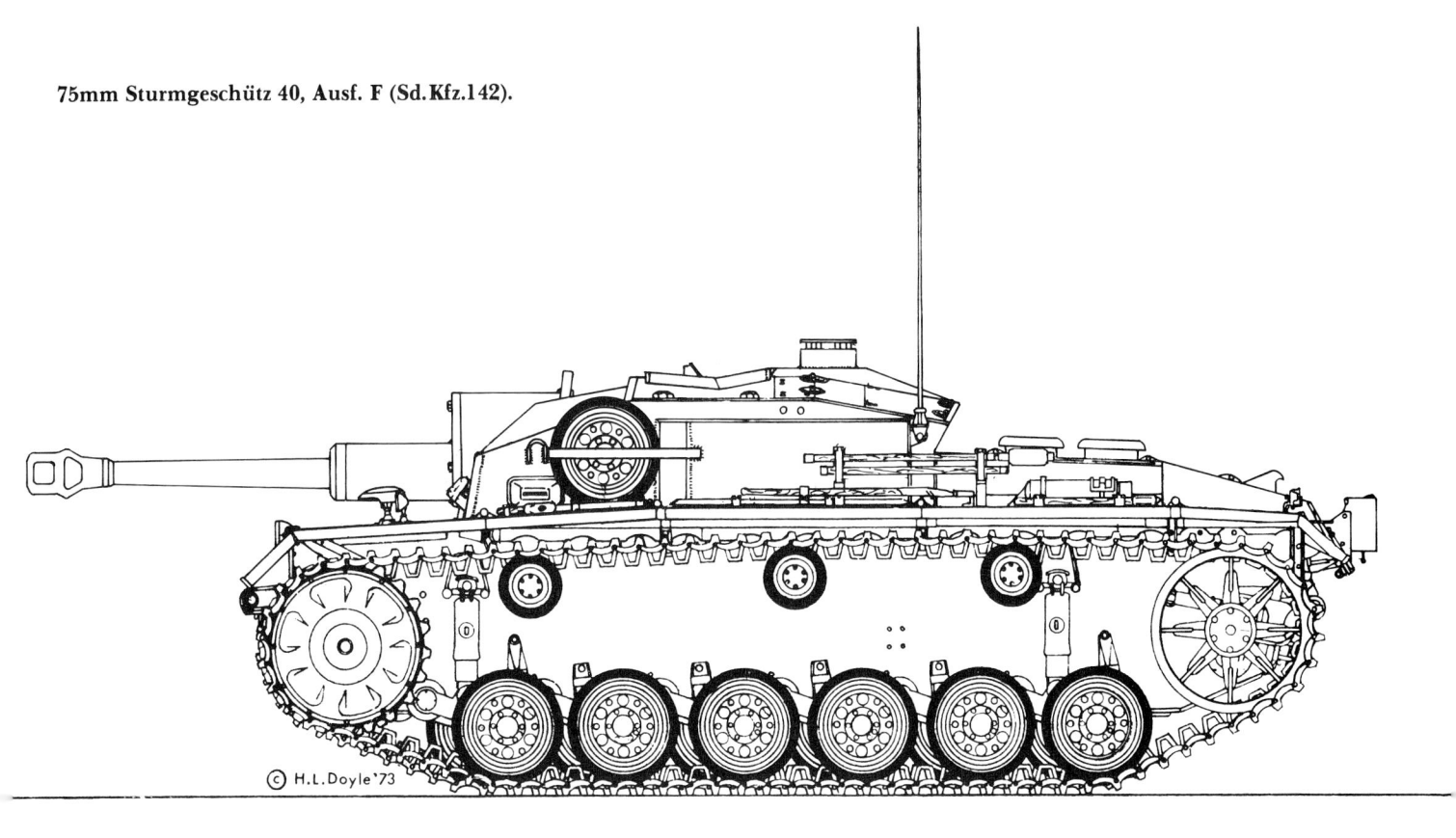

In September of 1941 Hitler ordered that the armor of the assault gun be strengthened without regard for increasing its weight or decreasing its speed. In addition, these vehicles were to have a gun with a long barrel and a higher velocity. This development was inaugurated in Document OKW 002205/41 gKdos, of September 28, 1941, and the chassis modifications were entrusted to the Daimler-Benz AG. The Rheinmetall-Borsig AG was made responsible for the gun and superstructure. In February of 1942, three test units with armor, and now carrying the 75mm Sturmkanone 40 L/43 as their primary armament, were finished. The basic armor remained unchanged for the time being. The targeting apparatus included one Sfl Z.F. 1a and one Rbl. F 36. The ammunition supply was 44 rounds. One Fu 16 or 15 radio was also included.

The vehicle with the 75mm Sturmkanone 40 L/43 was displayed to Hitler on March 31, 1942. On May 13 of that year Hitler stated that he agreed to a monthly production of 100 assault guns, and that the production of the Panzer III would thus be reduced to some 190 units. He also called for investigation of the possibility of strengthening the frontal armor of these vehicles to 80 mm, with this thickness to be attained first by welding on a pre-armor plate. The increased weight of some 450 kp was to be accepted. Initially there were problems with the new Sturmkanone 40, as it tended to jam after the first shot.

The monthly production increase to 100 units was confirmed on June 4, 1942, and the strengthening of the frontal armor to 80 mm was to be done before the middle of July.

The situation on all fronts required the immediate use of the 75mm Sturmgeschütz 40, Ausf. F (Sd.Kfz.142) which was then in production. With 50 mm basic armor, its fighting weight amounted to 21.6 tons. The entire production of these vehicles in 1942 numbered 828 units.

The firm of ALKETT led the way in producing these vehicles, and its production statistics were as follows:

| Month | Panzer III | Sturmgeschütz III |
|---|---|---|
| December 1941 | 25 | 40 |
| January 1942 | 28 | 40 |
| February 1942 | 35 | 35 |
| March 1942 | 37 | 30 |
| April 1942 | 37 | 25 |
| May 1942 | 42 | 25 |
| June 1942 | 45 | 25 |

**A Sturmgeschütz 40 Ausf. F of the "Grossdeutschland" Division in Russia. The angular mantelet armor was typical of this vehicle. As of the 120th unit, the 75mm Sturmkanone 40 L/48 was installed on the Ausf. F.**

The Altmärkische Kettenwerk GmbH had been established toward the end of 1937 as a branch of the Rheinmetall-Borsig AG firm for a monthly tank production (assembly) of 30 tanks of the Panzer II type at first. The first finished tanks of this type were produced during 1938.

These vehicles, some 300 of which were built at ALKETT, were replaced by the Panzer III and later also by the Sturmgeschütz, the latter remaining in production until the war ended.

The monthly output increased to some 50 vehicles during the first war year, and at the end of 1944 or the beginning of 1945 a maximum of some 380 Sturmgeschütz per month had been attained.

The increase to the maximum output was disturbed considerably by two serious bomb attacks on the factories, particularly the ALKETT I works in Berlin-Borsigwalde.

In the spring of 1944, most production was moved to the former Reich Railroad Maintenance Works of Albrechtshof in Falkensee. This facility had been taken over by DEMAG during the war, and this firm made some 30,000 square meters of the premises available to the ALKETT firm for Sturmgeschütz production.

In 1944 the following facilities were being used by the ALKETT firm:

Works I: Berlin-Borsigwalde
Works II: Berlin-Spandau, and
Works III: Berlin-Falkensee

In all, a surface area of some 50,000 to 60,000 square meters was available for Sturmgeschütz production. This firm built a total of 7700 armored vehicles.

By the autumn of 1943, complete vehicles (chassis and superstructure) were being completed or assembled at Borsigwalde. Forged components, cast steel parts, armor housings and other important components, such as engines, drive-train parts and the like, were delivered by supplier firms.

After the first major air raids in November of 1943, the Spandau facility was included for finishing work, and a greater measure of manufacturing and assembling parts was assigned to subcontractors in Berlin and elsewhere in the Reich. With great effort, the ALKETT III facility in Falkensee was prepared for a maximum monthly assembly capacity of 500 Sturmgeschütz. At the same time, the Spandau factory was prepared for the assembly of 500 superstructures and final assembly. In addition, this factory gained two hull-boring works for 250 hulls per month. Two additional hull-boring works of the same capacity were available at Falkensee.

At Borsigwalde, after the considerable destruction in November of 1943, only the swinging-arm line was rebuilt as an assembly line for 500 units per month. The gearbox facility in Borsigwalde, which was designed to turn out a maximum of 150 gearboxes per month, was rebuild after the bombing for 300 units (150 each of gearbox and steering-gear units) and equipped with the necessary machines and processing facilities. The rest of the gearboxes and steering gears needed for maximum production were produced by supplier firms.

A particular problem existed in the production of torsion bars. They were delivered by the firms of Röchling in Wetzlar, Hösch in Hohenlimburg and Dittmann-Neuhaus in Herbede. In addition, a small factory was set up in Berlin-Stralau.

The production of ball bearings caused great problems, above all because of the destruction of ball-bearing plants in Schweinfurt. It was foreseen that the new Panzerjäger 38 D, production of which was to begin at ALKETT in April of 1945, would have to be built almost exclusively with journal bearings.

Through the loss of the Continental firm of Hannover and other firms located in western Germany, the punctual supply of rubber tires for the road wheels and return rollers became a major problem. A similar situation existed for swinging arms, cranks and brake drums. Essentially, these difficulties were dealt with by extending and utilizing all available facilities and possibilities. In the end, the constant air raids had a catastrophic effect on the transportation situation, and even approximately sufficient supplying became impossible.

The firm of ALKETT held a special position in tank manufacturing, as compared to the other tank-building firms. On the basis of the firm's years of experience in tank construction and their available first-class personnel, the

**75mm Sturmgeschütz 40, Ausf. F/8 (Sd.Kfz.142/1).**

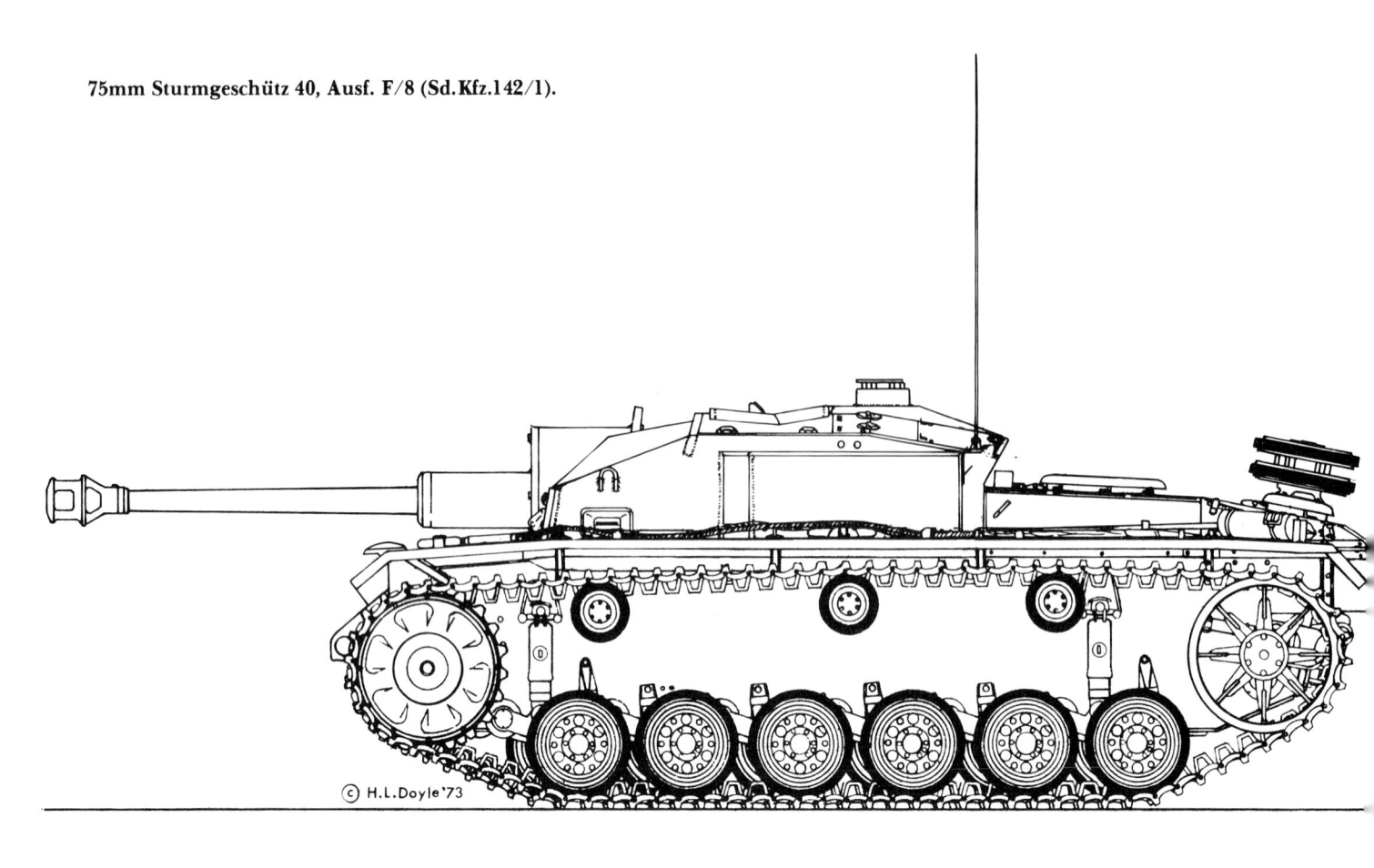

This version was designated "F/8" as it was based on the 8/ZW chassis. This picture shows a vehicle with armored side aprons.

firm was constantly called on by the Reich's minister for defense and war production to carry out important developmental tasks and to produce O series of new types that had been or were to be developed. ALKETT likewise had a significant share in the development of underwater and amphibious vehicles.

A look inside the Sturmgeschütz 40 shows the commander and gunner.

The loader loads the 75mm shell.

The loader also used the radio.

ALKETT was also a leader in the development in a so-called ultimate assault gun, for which the following minimum requirements had been established: 100 mm front armor, at least 40 to 50 mm side armor, 50 cm ground clearance, wider tracks, top speed 25 kph, lowest firing height, 75mm L/70 gun. The weight could thus rise to some 26 tons without causing concern.

Until the new assault gun could be put into service, production of the existing Sturmgeschütz was to be increased by all means. Investigations were conducted to see whether this could be attained by cutbacks in Panzer II and 38(t) self-propelled gun production.

The chassis numbers of the vehicles armed with the L/43 gun ran from 91 401 to 91 519. Starting with the 120th unit, the 75mm Sturmkanone 40 L/48 was installed. The Sturmkanone "75mm Sturmgeschütz 40", Type F/8 used the 8/ZW type chassis.

With the installation of the long "Sturmkanone 40" gun, the assault gun intended to be used against troops was turned into a tank destroyer, the most important task of which was to fight enemy tanks. For effective action against unarmored targets, an assault howitzer was created, differing from the assault gun only in having a 105mm leFH L/28 gun. This vehicle was shown to Hitler on October 2, 1942, and he particularly approved of the low firing height of 1.55 meters. A trial series of 12 vehicles was planned, six of them already being available at that time. Three more were to be finished by October 10, 1942, and the rest were to follow in about four weeks. Early in February of 1943, Hitler was informed that, beginning with 20 units in March and totaling 30 units a month as of May, assault guns with light field howitzers were to be produced. Hitler considered it important that at this point that hollow charge shells for this weapon would also be available. Of the 105mm Sturmhaubitze 42 (Sd.Kfz.142/2), there was first a Ausf. F, while the final version was Ausf. G. The 105mm assault howitzer originally had a muzzle brake to allow the use of added charges. According to Army Technical V.-Blatt 1944, No. 635, the muzzle brake of this weapon was eliminated in new construction. Already available Sturmhaubitze 42 kept them, but if they became unusable or no replacements were available, the muzzle brake of the leFH 18 M or leFH 18/40 howitzer

The whole crew helped to load ammunition aboard the vehicle.

The engine was often started by a hand-crank starter. It was not always an easy job.

The roof had wide locking hatches to make supplying easier.

could be used.

On the basis of front-line experience with the Sturmhaubitze, Speer suggested in mid-October of 1943 that the percentage of these vehicles in increased production be reconsidered. The Sturmhaubitze had become more and more unpopular to their crews in the course of their use, since they could not avoid tank battles but their guns were only usable in certain conditions. The need for tank destroyers toward the war's end pushed the assault howitzer completely out of the picture, and before the war ended it was taken out of production.

**The Maybach engines ran reliably, but replacing it was no rarity for members of the Panzer troops.**

A "Saukopf" ("pig's-head") mantelet for the 105mm Sturmhaubitze was also developed, but was installed in only some vehicles.

The Ausf. G of the "Sturmgeschütz 40" (Sd.Kfz.142/1) was given a widened superstructure, in which the hatch by the gun commander's position was replaced by a commander's cupola with periscopes. This cupola ring was originally able to turn 360 degrees but had to remain in a fixed position in later versions because of a shortage of ball bearings. The overall height of the vehicle had increased to 2160 mm. Later versions were fitted with additional shot deflectors on the hull roof, mostly in front of the commander's cupola; at first these were fitted by the troops, but later they were installed by the manufacturers. The driver's optics had been eliminated, as had the viewing ports on the sides. As of the "8/ZW" type of the vehicle, the towing hooks attached to the front and back with bolts were eliminated; they were replaced by extended hull sides drilled for shackles. The first Ausf. G vehicles had front armor 50 + 30 mm thick, while 30 mm plates were still used on the sides. In the end, the Sturmgeschütz 40 appeared with 80mm bow plates.

During the course of production, steel return rollers were also installed in order to limit the raw-material need for rubber. At the front of the upper bow plate, individual barrel travel locks were installed to which the gun-barrel was attached while marching. There were two different types of them. The gross weight had increased to 23.9 tons with 54 rounds of ammunition aboard. It must be mentioned that the factory-installed ammuition racks were removed by the troops in many cases, in order to allow a considerably larger number of shells to be carried. The price per vehicle was 82,500 Reichsmark.

The radio equipment was suited to the type of action. When the assault gun was used as a tank, either an Fu 5 or an Fu 5 plus an Fu 2 would be used. Vehicles that were used by assault-gun units had the following equipment: Fu 8, Fu 16 and Fu 15, or Fu 16 and Fu 15, or only an Fu 16.

Simultaneously with the development of the Panzerkampfwagen III, the Sturmgeschütz were also fitted with the obligatory aprons, originally fixed in place, but later free-swinging. The headlights were removable, and the vehicles themselves were treated with Zimmerit anti-

The Sturmgeschütz wrote their own unforgettable history as the reliable comrades of the infantry. They helped again and again against the enemy tanks that attacked in great numbers.

An L/48 gun with additional 30 mm armor plate on the front. This picture shows such a Ausf. F/8 vehicle.

The Sturmhaubitze with a 105mm light field howitzer served to attack unarmored targets. Except for its primary weapon, this vehicle was identical to the Sturmgeschütz 40 Ausf. F.

The 105mm Sturmhaubitze barrel was mounted in an angular cradle. The ammunition storage racks in this vehicle were adapted to fit the larger ammunition.

This picture shows the right side of the vehicle with the gun barrel depressed.

105mm Sturmhaubitze 42, Ausf. F (Sd.Kfz.142/2).

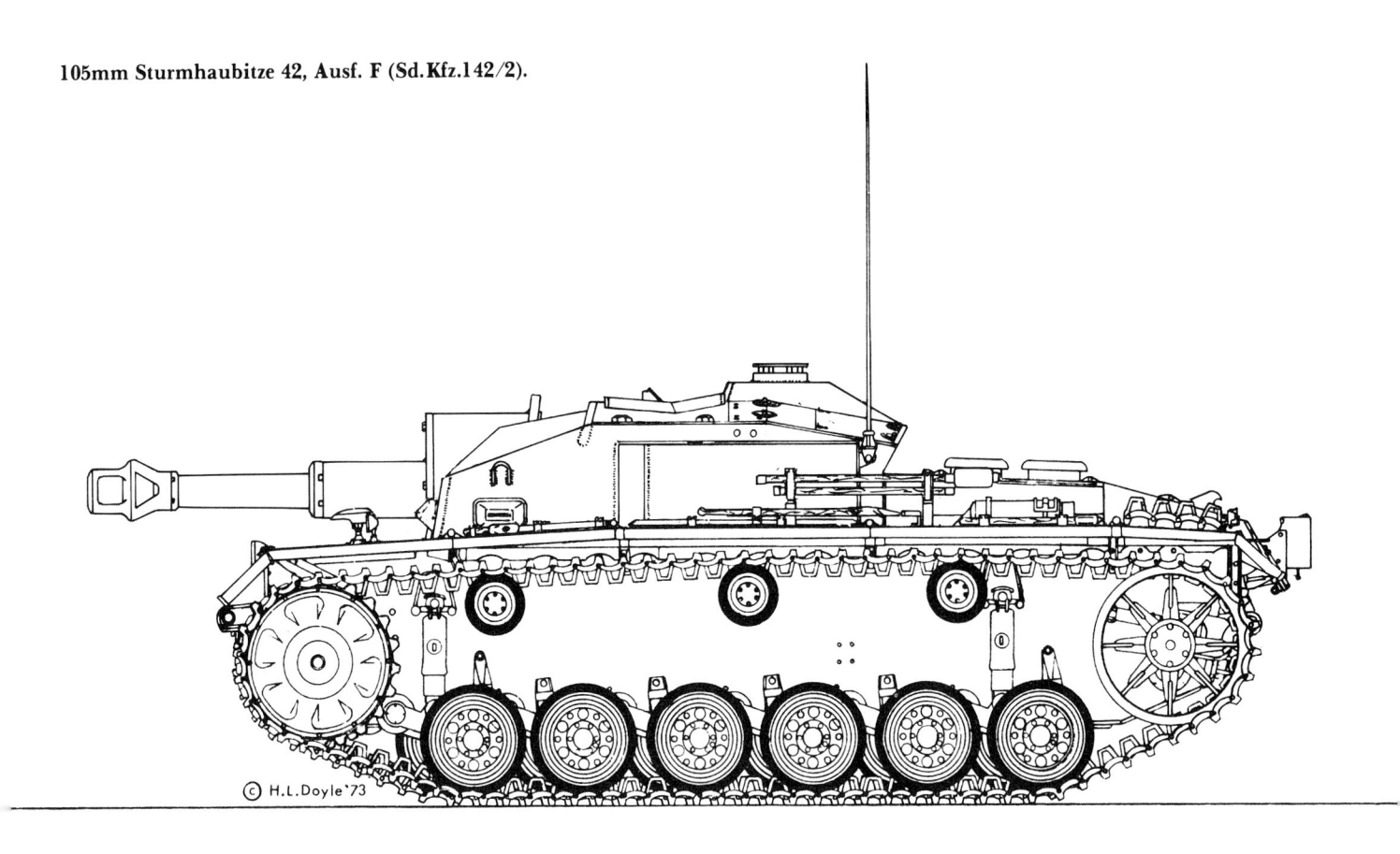

The barrel elevation allowed firing on targets under cover. These vehicles were of only limited use against tanks.

The Ausf. G version of the Sturmgeschütz 40 appeared in 1943 and had a wider superstructure. The gun commander now had a commander's cupola at his disposal.

This picture shows the Ausf. G from the front.

75mm Sturmgeschütz 40, Ausf. G (Sd.Kfz.142/1).

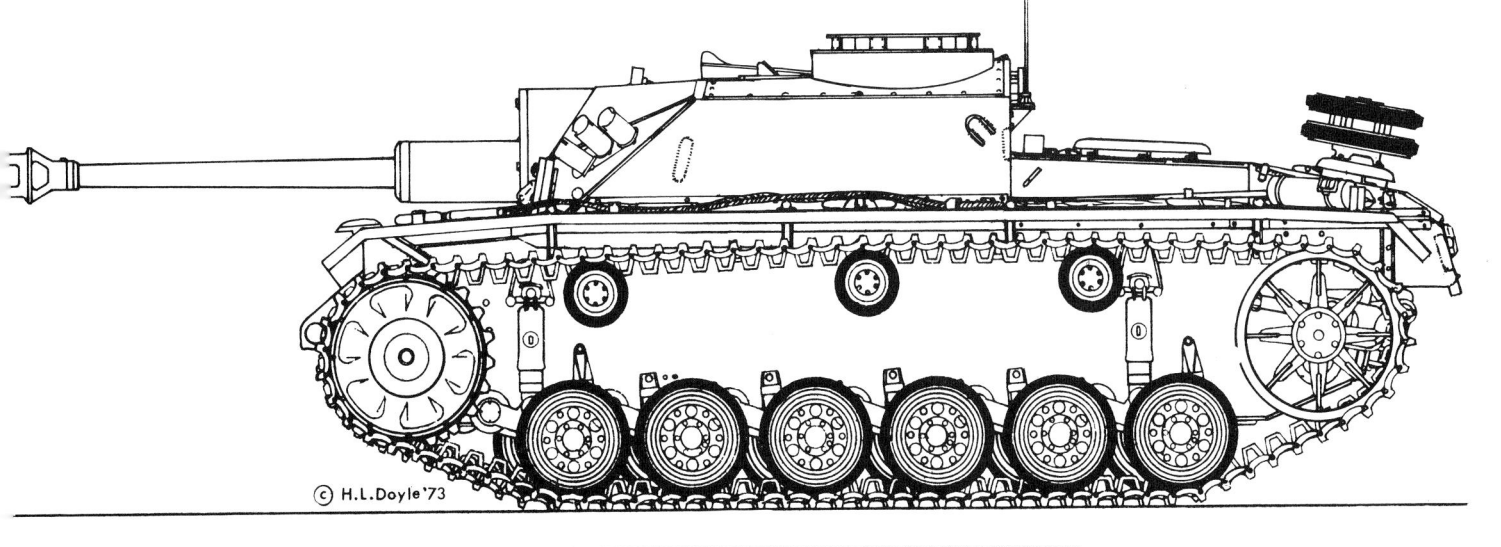

The production of these important vehicles was now carried on at full speed in various assembly factories. This picture shows the acceptance of complete Sturmgeschütz Ausf. G and howitzers.

111

Crew sleds in the Russian manner were tested as a means of waging mobile winter warfare. This picture shows such a sled hitched to a Sturmgeschütz 40 Ausf. G.

The Sturmhaubitze 42 Ausf. G was originally equipped with a muzzle brake.

105mm Sturmhaubitze 42, Ausf. G (Sd.Kfz.142/2).

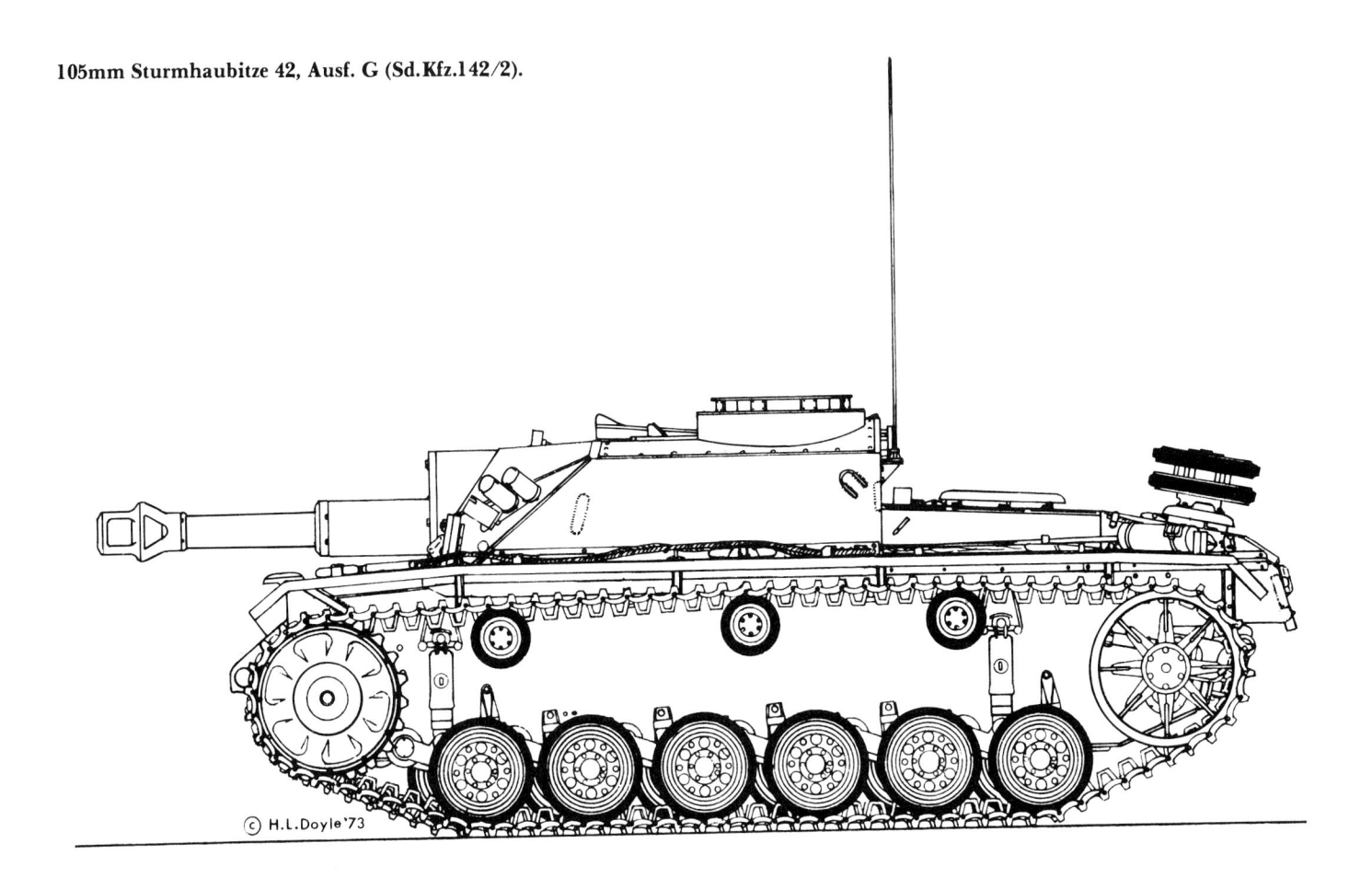

magnetic protection. Along with the smoke-laying apparatus attached to the body, the asymmetrically widened "Ostkette" tracks were also used on account of the poor ground conditions in Russia.

The ventilator mounted on the roof as of Ausf. F was now moved to the back of the superstructure.

In this form the "Sturmgeschütz 40", urgently needed at the front, was built in growing numbers until the end of the war. The German Army's complete supply of Sturmgeschütz numbered 716 on July 1, 1941, and 623 on April 1, 1942. By June 1943 the monthly output was supposed to be increased to 120 units. Sturmgeschütz production was less costly and time-consuming than tank construction, and reached its highest point in 1944. In 1943 3245 of them were built, and in 1944 a total of 5751 assault guns were produced, including Panzerjäger vehicles on the Panzer IV chassis.

**While this vehicle was in production, the muzzle brake was eliminated. The use of added charges was no longer possible.**

Of these, 4850 were Sturmgeschütz with the 75mm gun, while 901 carried the 105mm howitzer. Another 145 of the Assault Gun 40 were built in 1945.

In order to produce these quantities, the MIAG Amme works in Braunschweig was utilized for assault-gun production as of 1943, in addition to the ALKETT works. MIAG alone attained the following production figures:

| Month | 1943 | 1944 | 1945 |
|---|---|---|---|
| January | — | 120 | 71 |
| February | 42 | 137 | 37 |
| March | 85 | 120 | 15 |
| April | 122 | 100 | — |
| May | 136 | 80 | — |
| June | 120 | 145 | — |
| July | 120 | 135 | — |
| August | 120 | 80 | — |
| September | 140 | 100 | — |
| October | 138 | 72 | — |
| November | 130 | 110 | — |
| December | 117 | 91 | — |

A "Saukopf" ("pig's-head") mount was also introduced for the 105mm Sturmhaubitze, but it was actually installed in only some vehicles. This picture shows such a vehicle with the new mount and muzzle brake.

Chassis numbers for Sturmgeschütz with long barrels were numbered, among others, from 91 401 to 93 850 and from 95 116 to 95 229.

The V-shaped, angular, welded gun-cradle armor installed along with the Sturmkanone 40 L/43 or L/48 was replaced in mid-1943 by cast-steel armor in shot-deflecting form, the so-called "Saukopf" ("pig's-head") mantelet. The 30 mm armor added to the basic armor in the front was partly screwed on and partly welded. Originally there was no mount on the vehicle for the MG 34 (later MG 42) that was carried for self-defense. In the spring of 1943 a machine-gun shield was introduced that gave added protection to the loader. The shield was attached to the roof and could be folded. The front half of the loader's visor held the protective shield in its upright position.

Toward the end of 1944, newly-built Sturmgeschütz were equipped with a machine-gun mount that had been developed originally for the Jagdpanzer 38(t) "Hetzer." It was designed so that it could hold a machine gun which could be operated from inside the fighting compartment. A periscope had to be used, since the loader's eyes were some 46 cm below the weapon. The periscope had threefold magnification and an eight-degree field of vision. The machine gun, located behind tube deflector panels, could be turned 360 degrees. Its installation required a change to the hinges of the loader's hatch. The

105mm Sturmhaubitze 42, Ausf. G (Sd.Kfz.142/2), the version without a muzzle brake.

Because of material shortages during production, steel return rollers were used. It is obvious that some of this captured vehicle's forward torsion bars are broken.

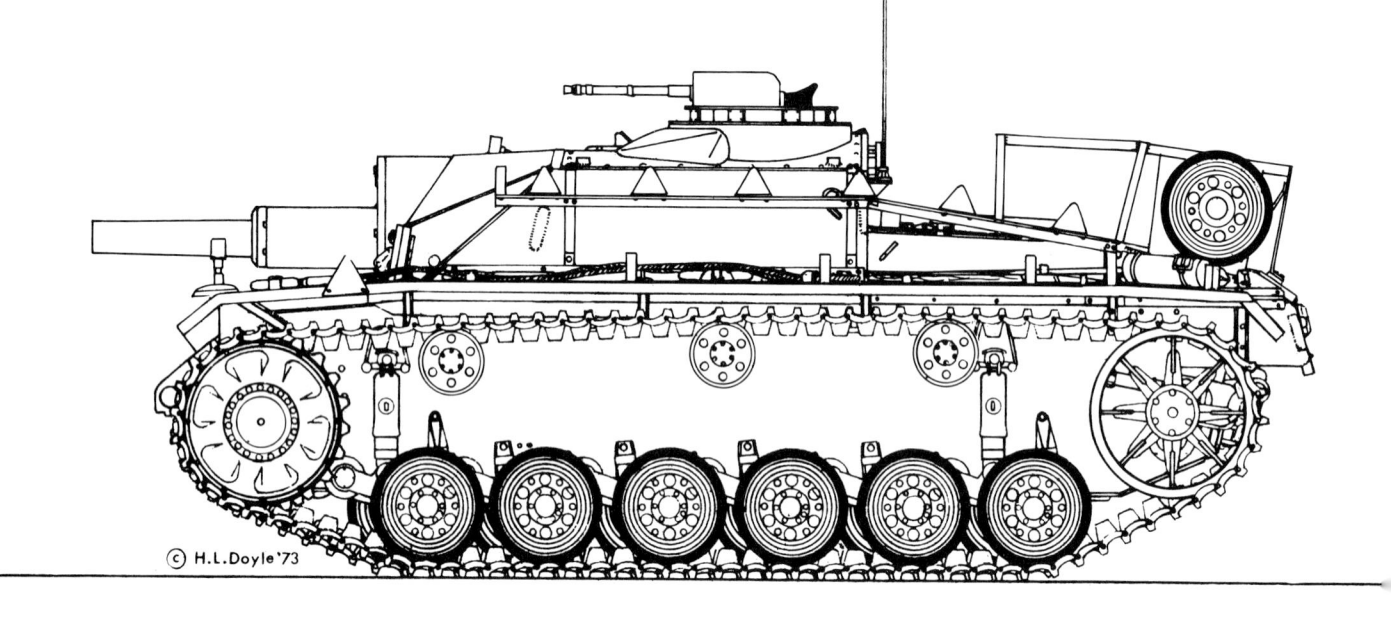

105mm Sturmhaubitze 42, Ausf. G (Sd.Kfz.142/2), final version.

Steel road wheels made by the Deutsche Eisenwerke were also used experimentally on the Sturmgeschütz 40. This picture shows the spare wheels of this type carried on the back of the vehicle. The license plate on the right fender, marking it as an experimental vehicle, was used during testing. The full-track armored vehicles of the Wehrmacht generally did not carry license plates.

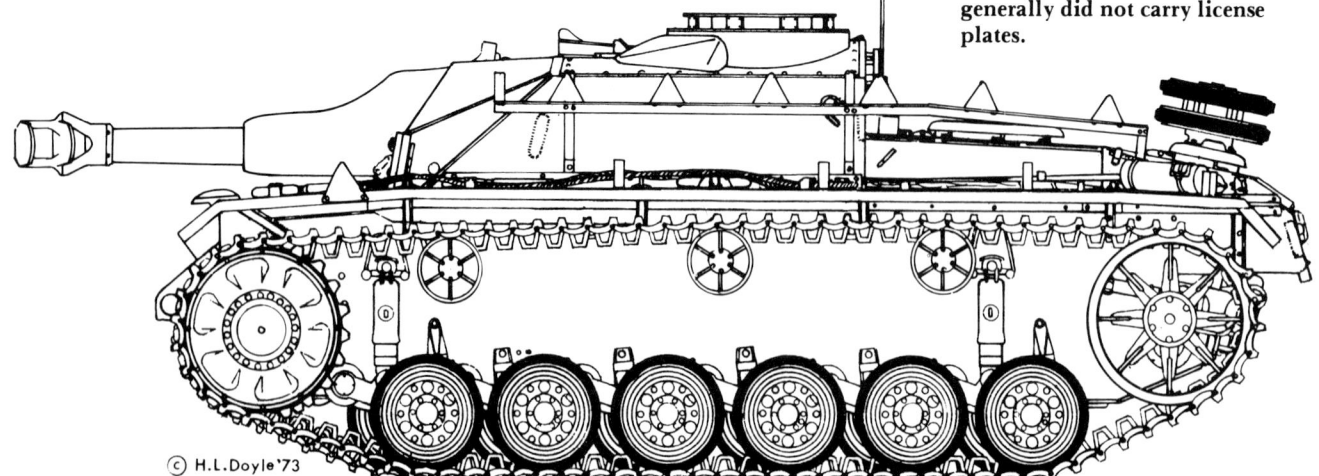

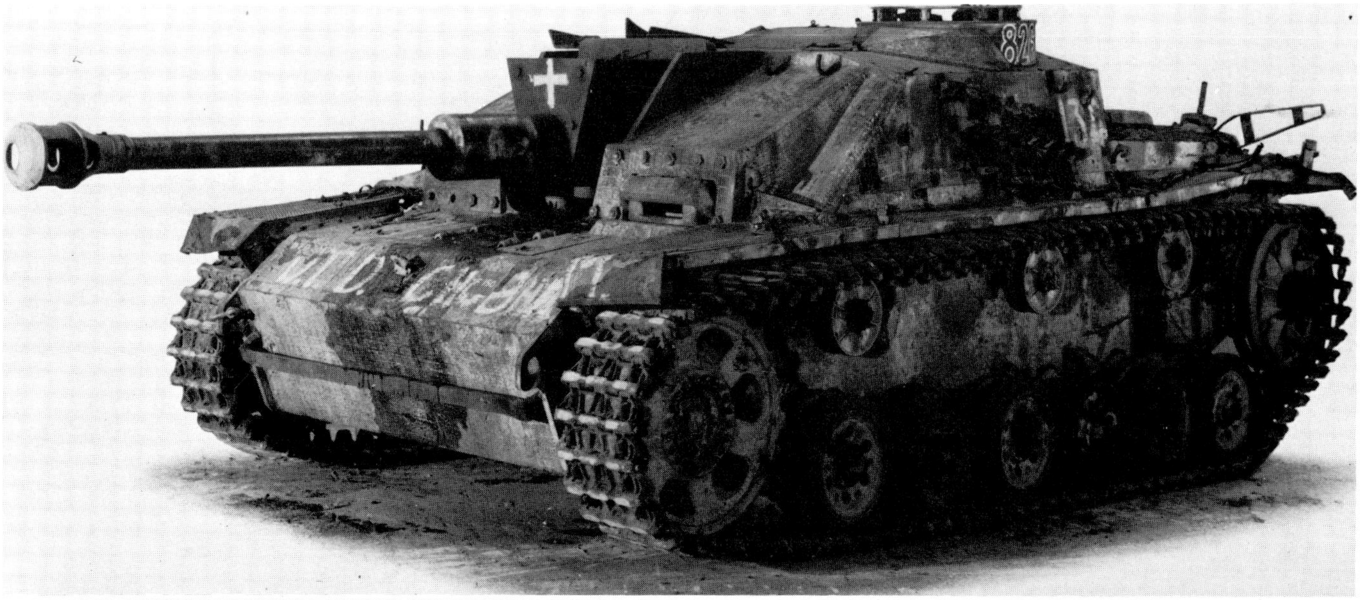

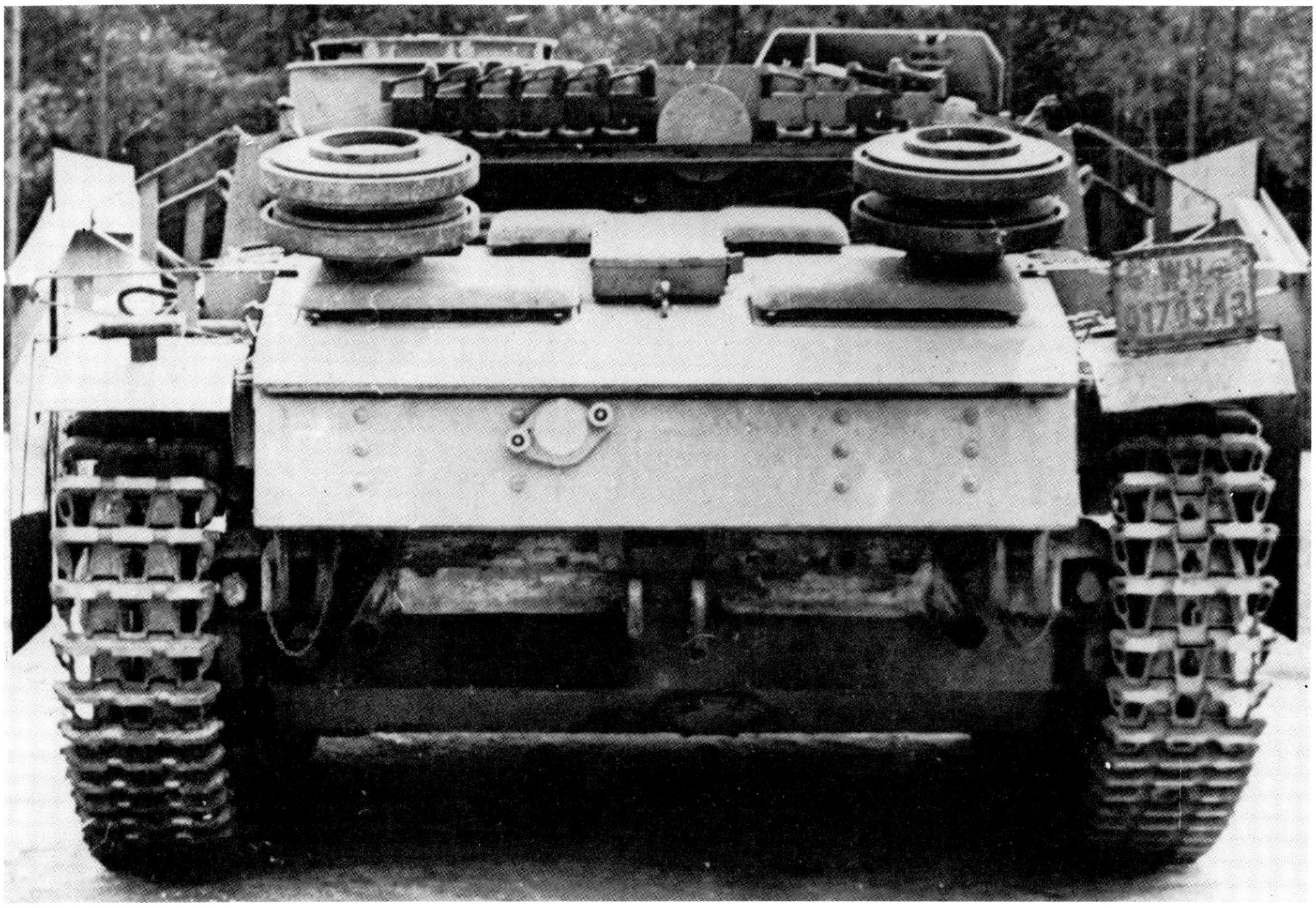

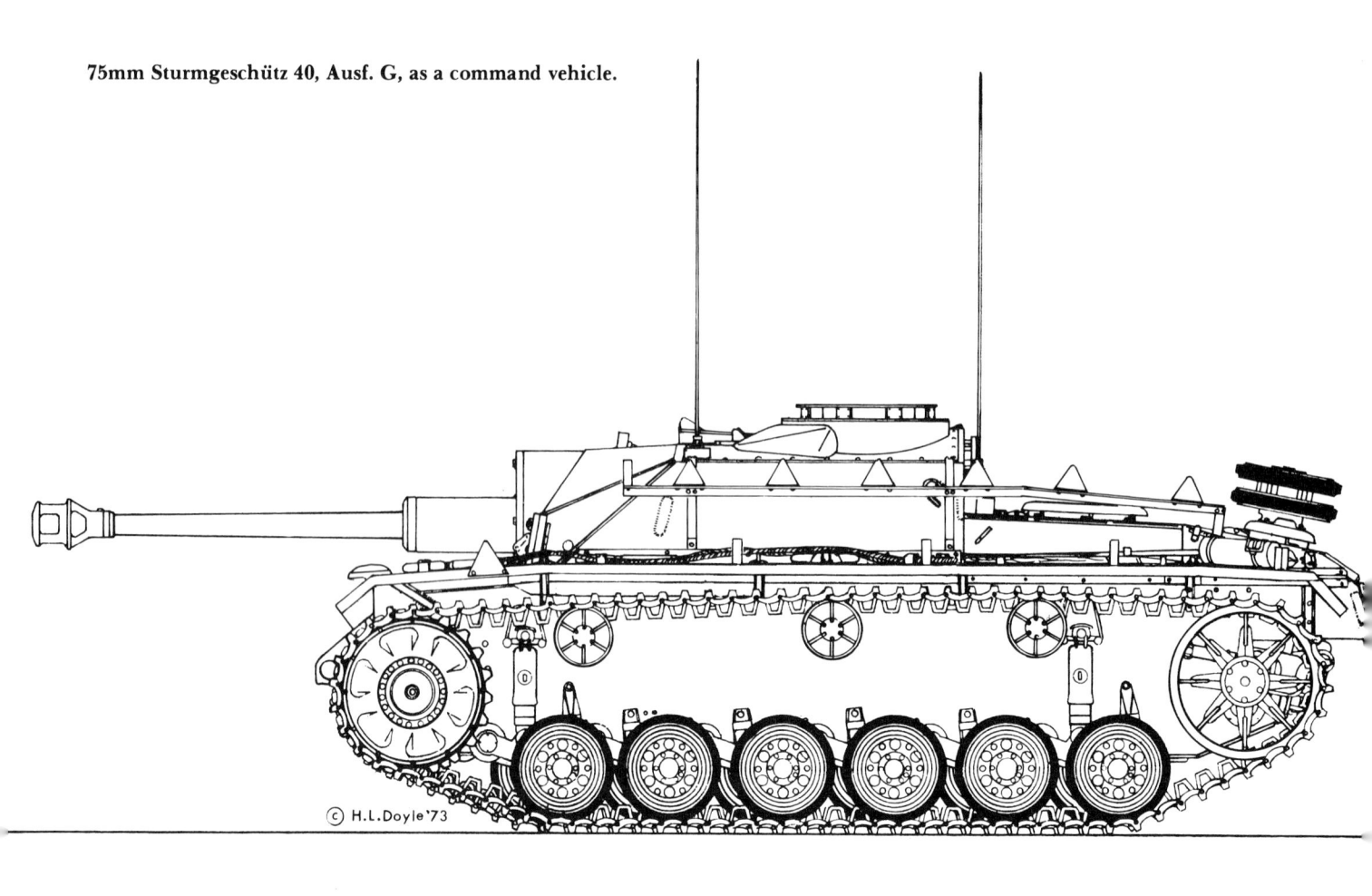

**75mm Sturmgeschütz 40, Ausf. G, as a command vehicle.**

© H.L.Doyle '73

The angular gun-mantle armor was a typical identifying mark of all assault-gun versions for some time. As of mid-1943 it was partly replaced by a cast steel fitting in shot-deflecting form, the so-called "Saukopf."

The loader had a machine gun available for self-defense and for firing on unarmored targets. This weapon was now mounted behind a folding protective shield. This picture shows the shield attached to the front cover of the loader's hatch.

All Sturmgeschütz had a four-man crew. The "Zimmerit" protective covering to prevent the attachment of magnetic mines is easy to see.

A camouflaged Sturmgeschütz in action on the western front. The picture shows a modified type of apron attachment. The concrete shot deflector in front of the cupola can also be seen.

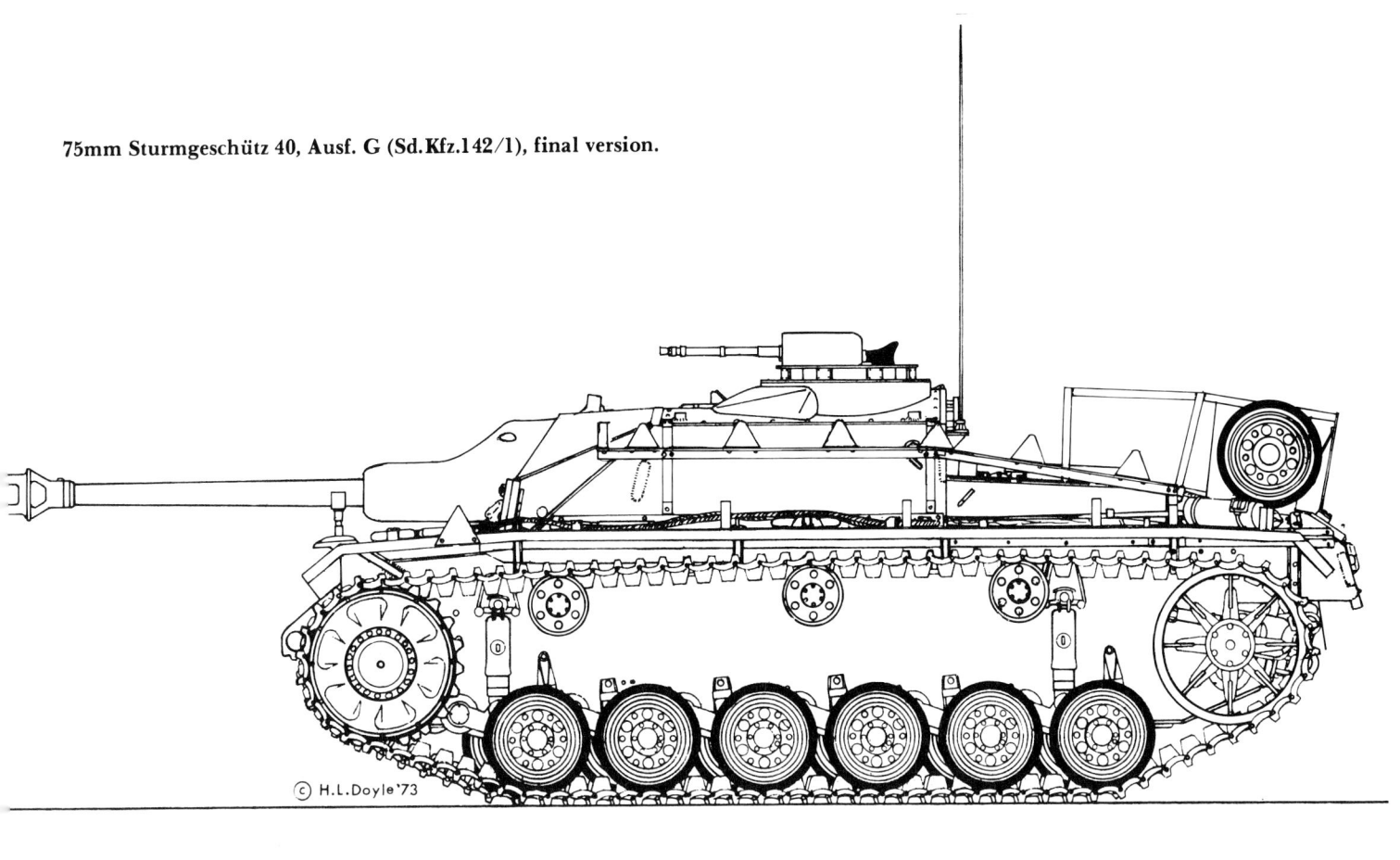

75mm Sturmgeschütz 40, Ausf. G (Sd.Kfz.142/1), final version.

This Sturmgeschütz 40 Ausf. G is in the Finnish Army Museum today. The cutaway wall of the fighting compartment allows three members of the crew, the driver, gunner and gun leader, to be seen.

Several associated armies, including that of Finland, used German Sturmgeschütz during World War II. Finland did not muster out the last of its Sturmgeschütz until late 1966.

halves of the visor hatch now opened toward the sides of the vehicle.

The desire for a coaxial machine gun was also fulfilled toward the end of the war; attachments for a machine gun were added to the left of the gun-barrel deflector and a steel tube was fitted into the gun-mantle armor. This machine gun was operated by the aiming NCO through the use of the gun's aiming optics and made it possible to fire on unarmored objects.

As a result of air-raid damage, Sturmgeschütz production decreased rapidly in November 1943 and even more so in December. The monthly quota ordered by Hitler in January of 1944 could be met and exceeded only in March of that year. In that month, 549 units were delivered.

The fighting value of the Sturmgeschütz and Panzerjäger can be seen from, among other things, the fact that, according to army reports, 22 to 32 percent of enemy tanks shot down were achieved by these vehicles from August to October 1944. Although the number of tanks in service was greater, they achieved only a score of 22 to 26 percent.

### Variants

The battles around Stalingrad proved the necessity of having a heavily armored vehicle available that was capable of knocking down buildings with a heavy gun that could fire shells that acted like mines. High muzzle velocities and long ranges were of little value here, but good armor was decisive. Hitler ordered that everything be done to prepare 12 such vehicles in two weeks if possible,

A Sturminfanteriegeschütz 33 in action in Russia, 1943. The picture shows a destroyed "Stalin Organ" beside the vehicle, as well as a ZgKw and a Zündapp KS750 motorcycle with sidecar.

but at least six to start with. If the installation of the 150mm sIG 33 (heavy infantry gun) in the Panzer III or IV turret should not be possible, then efforts should be made to install this gun in the Sturmgeschütz vehicle. Early in November 1942, twelve of these vehicles were ready for service at Stalingrad. By the end of November, 24 of them were to be finished. This development, consisting of mounting a heavy infantry gun on a Sturmgeschütz chassis, was introduced as the result of a comment in a report dated July 16, 1941.

This "Sturm-Infanteriegeschütz 33" on the Sturmgeschütz chassis had a five-man crew, a supply of 30 rounds of ammunition, and a fighting weight of 21 tons. The body, enclosed on all sides, was protected by 80 mm armor on the front, 50 mm on the sides and 15 mm in back, plus 10 mm plates on the roof.

During the course of further development of this vehicle, the "Brummbär" armored assault vehicle (Sd.Kfz.166), built on the Panzer IV chassis, was created. The Sturminfanteriegeschütz 33 saw service with the 9th Company of Panzer Regiment 201 in the summer of 1943.

Plans were made in 1943 for the development of a fast-moving, light armored vehicle with heavy armament. For the sake of quick series production, it was to be made chiefly of components of the Panzerkampfwagen III developed by Daimler-Benz. This project was canceled on November 22, 1943.

On October 20, 1943 an experimental model of a Panzer III, armed with the 75mm KwK L/24 and with provisions for running on railroad tracks, was displayed at the Arys troop training center. This rebuilding project was developed by the Austrian Saurer Works in Vienna. The

**Sturm-Infanteriegeschütz 33.**

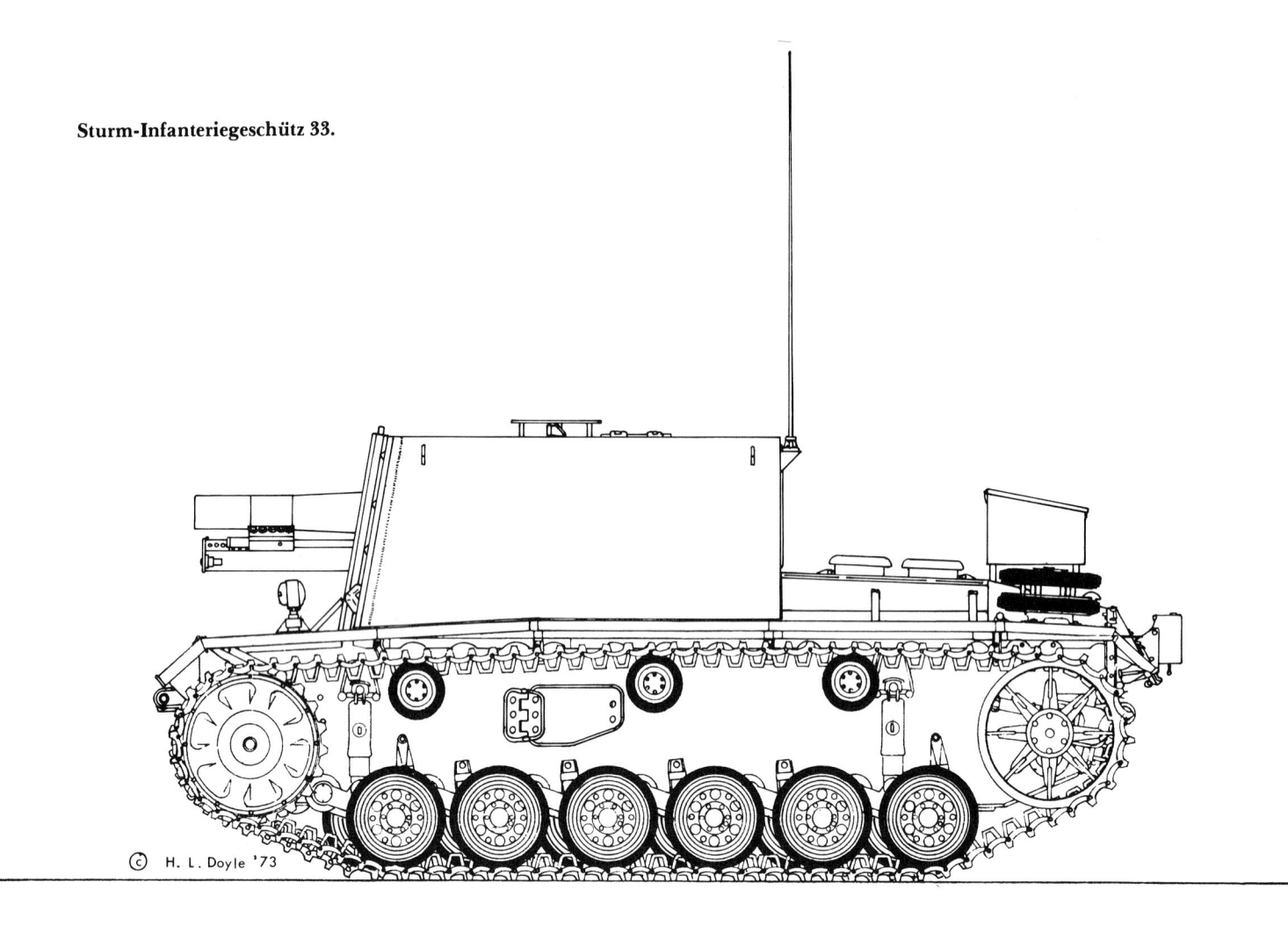

vehicle was equipped with retractable railroad wheels whose hubs were driven by the vehicle's engine via four spindles. Speeds of up to 100 kph on rails were attained. This vehicle was to be used for protecting railroad trains in areas of partisan activity.

An experimental vehicle was made of Panzer III components for use as a minesweeping tank. The running gear was adapted to provide considerable ground clearance by lengthening the swinging arms.

Armored engineer units received a Panzer III variant with a fast bridging device in 1943; this made it possible to cross small obstacles in the country. These vehicles were made without a rotating turret and had their engineering equipment on top of the hull for quick deployment.

So-called "Schlepper III" towing tractors built for transport tasks, and fitted with a wooden loading bed, also appeared for service on the eastern front. Most of these vehicles were fitted with "Ostkette."

Panzer III vehicles without rotating turrets were also used in great numbers by the troops as ammunition supply vehicles.

Several maintenance units were supplied with similar vehicles to serve armored units. The turning circle of the turret was usually covered with a wire frame with a tent canvas stretched over it.

One of the most frequent variants of the Panzer III was a Bergepanzer (recovery vehicle) that was based on a normal Panzer III chassis without any great changes. The equipment needed to tow vehicles was carried in this tractor. At the beginning of 1944 it was reported that 15 of the Panzer III vehicles to be finished in March and 30 each in April and May were to be rebuilt as recovery tractors.

Two more Sturminfanteriegeschütz 33 of Panzer Regiment 201. The picture shows the radio-operator's bow machine gun, used for close combat. Next to the first vehicle is a light "standard" passenger vehicle.

A Panzerkampfwagen III rebuilt by the Saurer-Werke in Austria so as to run on railroad tracks. The vehicle was intended for use in protecting trains from partisan attacks.

Only prototypes of the mine-sweeping Minenräumpanzer were built. The considerably higher ground clearance was supposed to provide the crew with better protection against exploding mines. The weight of the vehicle was about 20 tons.

To increase the number of recovery vehicles, these rebuilt Bergepanzer III units were used. They were fitted with a wooden body. The so-called winter Kette are fitted.

Rebuilt Panzer III vehicles were also used in great numbers as armored ammunition carrier Munitionspanzer III. This picture shows one of those vehicles supplying ammunition to a Tiger tank unit.

In May of 1944 Hitler described the armored road-clearing device, developed at the urging of Gauleiter Jordan and used to remove debris resulting from air raids, as particularly valuable and ordered that ten Panzer III or IV chassis then undergoing repairs be devoted to this purpose.

Plans were also made for the following self-propelled guns that were to utilize a Panzer III chassis: Gerät 814 as a 634/6 gun car (Gw) for the 105mm LeFH 18/40/6 Sf; Gerät 815, the Gw 634/4 (K 8 F 43/2 Sf); and Gerät 816, the Gw 634/3 (210mm GrW Sfl.). No further information on these vehicles has come to light at this time.

In a few cases, variants of the Sturmgeschütz appeared without primary armament as ammunition carriers. In addition, according to a memo dated November 1942, ten assault guns were to be equipped with flamethrowers.

Captured Panzer III vehicles were used, particularly by the Red Army, for training, and some of them were also used by the "Free Germany National Committee." In Africa, units of the "Free Polish Forces" used such vehicles for training. In 1943 the Soviets rebuilt Panzer III vehicles for use as self-propelled guns, equipping them with the 7.62 cm Pak (SU 762) gun. Such a vehicle was captured by the 23rd Panzer Division in the autumn of 1943 and saw service with Panzerjäger Battalion 128.

One of these vehicles can still be seen today as a monument at Sarny in the Ukraine. The vehicle was retrieved from a river into which it had broken through the ice during the winter.

Some years after World War II, several Panzer III tanks were still seeing service in Norway and Denmark.

Of the Reich's wartime allies, Finland obtained thirty Sturmgeschütz 40 units in 1943, at a price of some 52,000 Reichsmark apiece. Another 25 of these vehicles were ordered in 1944, this time at a price of about 32,000 Reichsmark each. In July of 1944 a monthly delivery of 15 of these vehicles was still approved. At the same time, two FAMO 18-ton towing tractors and various repair-shop equipment were delivered to Finland.

Supply became an ever-greater problem, and in order to keep other vehicles ready for action, seven Sturmgeschütz were dismantled for parts in 1944.

Panzer III chassis were also made available to the tank repair units. After the turret was removed, a removable wire frame was attached and covered with a tent canvas to protect the crew from wind and weather.

Most of the mustered-out Panzer III chassis, though, were used for tank recovery vehicles. They provided valuable service in towing broken-down armored vehicles. A large percentage of these vehicles came into being when the troops rebuilt them.

Armored engineer units were supplied with an armored engineer vehicle based on the **Panzer III** as of 1943. These vehicles carried engineering equipment and a fast bridgelaying device for spanning certain obstacles in the terrain. This picture shows such vehicles in action.

**Small numbers of Panzer III chassis with assault-gun bodies were used as ammunition carriers. This picture shows such a vehicle after being captured by American troops.**

**After the war, a number of Panzer III chassis were used, after their armored bodywork was removed, as railroad engines for supply work. The lack of locomotives compelled this practice.**

After the war, additional spare parts were purchased in 1948 from the stocks of supplies stored in Norway, and which included a complete Sturmgeschütz 40 and about twenty Panzer III chassis. Some of these vehicles were used for training purposes, while the others were stored as army reserve stocks. In 1960, though, all these vehicles were released for use in training; in all, 45 Sturmgeschütz 40 units were available. In December of 1966 these vehicles were all scrapped.

In the summer of 1944, some forty Sturmgeschütz 40 units and other equipment were delivered to Bulgaria, and a similar quantity went to Rumania. Finland, Bulgaria and Rumania used these vehicles against the German Wehrmacht in the autumn of 1944.

As already noted, 56 Panzer III units equipped with 50mm L/60 guns were delivered to Turkey in 1943.

Each of the well-known international armored vehicle museums now has several of these vehicles on display.

**Some Panzer III chassis were fitted with Russian superstructures. After a Russian 7.62 mm antitank gun was installed, they were known as "SU 762" tanks. One of these vehicles is still used as a monument in Sarny, in the Ukraine.**

A great number of Panzer III chassis fell into the hands of the Allies. Russia obtained the greatest share of them. They were used for training purposes. This picture shows Russian tank crews in German Panzer III vehicles.

In April of 1944 Hitler ordered that the final testing and the start of series production of the new, simplified running gear of a uniform chassis for Panzer III, IV and assault guns be carried out as quickly as possible. The uniform III/IV chassis will be described below.

The fact that, from 1935 to 1945, a total production of 15,350 ZW vehicles was attained speaks for the reliability of these vehicles. The Panzer III itself was a very modern tank in its day, and it proved itself very well in every way on all fronts.

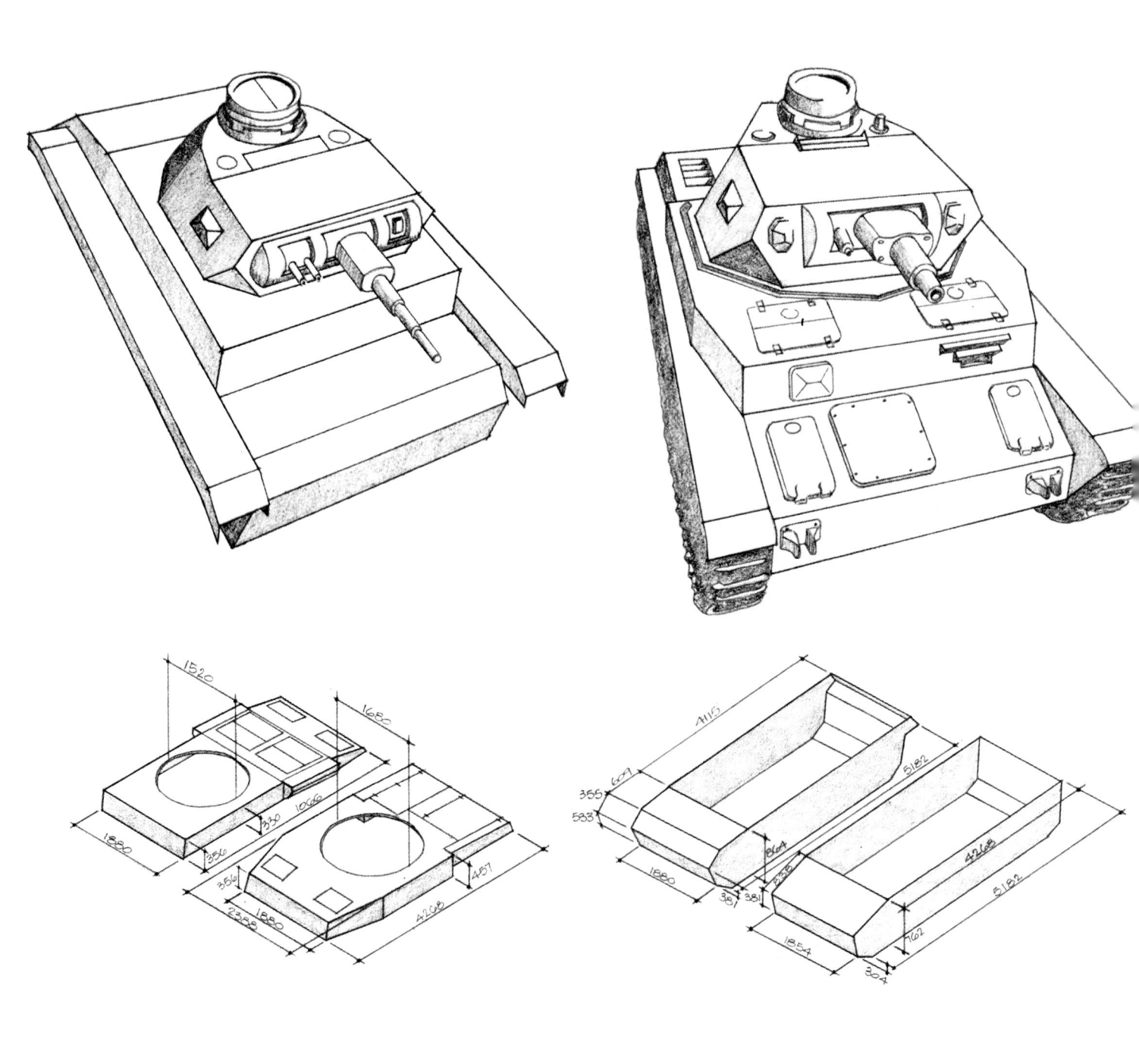

How little the "ZW" and "BW" vehicles actually differed can be seen in these drawings of their hulls, superstructures and dimensions. The idea of combining them to make a single vehicle was obvious.

# The Panzerkampfwagen III/IV

According to a statement made on September 6, 1941, a series of advances could be gained by making the Panzer III and IV more uniform. Thus the two vehicles became more and more similar in both their technical and tactical characteristics. The rearmament carried out frequently during the war also made their fighting value more uniform. Thus the idea necessarily emerged that only one battle tank could be built instead of two.

One external identifying mark of this Panzerkampfwagen III und IV n.A.) (neue Ausf.: new version) was the elimination of return rollers through the use of large-diameter road wheels in box arrangement, similar to those of the halftrack towing tractors. The request for large-diameter road wheels was tested in particular by the testing department of the Krupp firm (Egeln department) in Essen, which worked out such designs for both the Panzer III and Panzer IV. The existing suspension system was preserved to a great degree, but a new, wider track was given consideration. Several experimental versions of the Panzer III chassis were actually built and utilized, chiefly for training purposes, at various troop training facilities. The experience of FAMO in this area was given much consideration. A decrease in wheel pressure was anticipated, as were both a higher tolerance to wear and a decrease in ground pressure. One of the test vehicles built by Krupp later served as the basis for the Panzer IV c self-propelled gun carriage.

The uniformity of the two vehicles (newly designated Panzerkampfwagen III/IV) was expected to offer the following advantages:
Less work for the design office,
Less work in testing,
Less work in terms of test-series production,

Full uniformity of all parts (except armament), resulting in higher production, simplified supplying, and simplified and uniform training.

Greater flexibility in production. If need be, Panzer III tanks could be produced instead of Panzer IV, or vice versa, in a relatively short time. This would make for greater flexibility in organization. Considerably increased supplies of ammunition for the Panzer III (some 40 percent more than before, although the future shells would be larger than those previously used).

The possibility of installing a stronger armor-piercing weapon than the 50mm KwK L/60 then in development in the turret during the course of development, such as the 0725 weapon with conical barrel.

The opportunity to make the turret of the Panzer III rotate hydraulically, resulting in less of a burden for the crew and a higher rate of fire.

As for armor protection, the two vehicles were to be covered as follows:

| Vehicle | Front | Side | Rear |
| --- | --- | --- | --- |
| Panzer III n.Ausf. | 50mm | 50mm | 50mm |
| Panzer IV n.Ausf. | 50mm | 40mm | 40mm |

In order to keep the gross weight of both vehicles at a maximum of 23.5 tons, the Panzer III n.Ausf. was to have a light hull width of only 1.65 meters, as opposed to the 1.8 meters of the Panzer IV n.Ausf.

Then as before, though, the full uniformity of having just one type was striven for as a desirable goal. The only difference between the two tanks was to consist of the necessary divergence of their equipment. But this thoroughgoing uniformity could be attained only if, in using the hull (1.8 meters wide) of the Panzer IV n.Ausf., the side armor of the new-type Panzer III could be lightened somewhat to keep the weight down to 23.5 tons. The idea was that those surfaces of the sidewalls that were covered by the running gear, plus the rear, would have armor only 40 cm thick instead of the originally planned 50mm plate.

A letter from the Chief Rüst und BdE, Stab Rüstungs 11a, No. 2944/41 g, dated September 14, 1941, states that:

The uniformity of one type of Panzerkampfwagen III/IV is purposeful and is to be carried out (differences only in equipment).

The large-diameter in-line road wheels were an external identifying mark of the Panzerkampfwagen III/ZW40. Several of these vehicles were actually built.


Such vehicles were sent to troop training centers for use in training tank troops in close combat. The picture shows the attachment of a hollow-charge explosive charge to the engine compartment of a "ZW40."

After thorough winter testing, these vehicles were also used to pull the heavy sled trailers made for tracked vehicles.

The weakening of the armor on those surfaces of the hull sides that are already inherently covered by the running gear should not be undertaken. It would not be justifiable to avoid any increase of the 23.5-ton weight so fearfully. What with the Führer's order to add pre-armor to protect the front surfaces under any conditions, the previous maximum weight would have to be exceeded in any case. Thus the weight of 23.5 tons could not be retained anyway.

The Krupp-Gruson Works was actively engaged in this development and created, on the basis of the Panzer III/IV, the Panzer 604/9, formerly designated "Panzer IV long (A) 75mm Pak 43." The Panzer 604/10 was basically a Diesel vehicle. ALKETT delivered a new development with the designation of 604/11, which was also to be armed with the 75mm Pak 43 instead of the Gerät 21-559 (604/10). Another vehicle, the Device 561, a Panzerkampfwagen 604/15, was also mentioned.)

The following suggestions for gun carriages using the Panzer III/IV chassis were made:

Gerät 804, Geschützwagen III/IV, for leFH 18/40/2 Sf,
Gerät 807, Geschützwagen III/IV, for sFH 18/1 Sf, and
Gerät 812, Geschützwagen III/IV, for sFH 18/5 Sf.

ALKETT prepared the so-called new assault gun on this chassis, which was listed as Device 821, 75mm L/48 Panzerjäger III/IV on Panzer IV chassis. Krupp-Gruson produced designs for Gerät 824, a Panzer III/IV uniform vehicle as a battle tank. The last of these objectives that were never produced was Gerät 565, a 604/14 load carrier for the 30.5 cm mortar.

The "Preliminary Production Targets Program IV" of July 14, 1944 provided for manifold use of the uniform III/IV chassis. Because of steady increases in caliber, though, the previous 24-ton chassis had become unusable. New developments therefore called for weights of around 35 tons. For this uniform chassis the Deutsche Edelstahl AG delivered a body for a "light Panzerjäger III/IV" to MIAG in Braunschweig in September of 1944. It was never completed, though, because at this point the project had once again been given up by the Army Weapons Office. In the project mentioned above, this vehicle is listed as "75mm Sturmgeschütz III/IV L/70", and the firms of ALKETT and MIAG were to begin large-series production of it in November 1944, Krupp-Gruson in January, the Nibelungenwerke in February and Vomag in March of 1945. As of May 1945 the production quota was to be 800 units per month.

Plans were made by ALKETT to use the same vehicle, once production of the 105mm Sturmhaubitze 42 had ended in April of 1945, as the basis for a 105mm Sturmhaubitze III/IV. The monthly quota of these vehicles was to be 125 units.

The firm of Stahlindustrie GmbH was given the task of supplying the army with self-propelled artillery guns on this chassis. Plans were made for a "heavy armored howitzer" and an accompanying "ammunition vehicle for the heavy armored howitzer." The production quota was to be 25 howitzers and four ammunition carriers per month as of May 1945.

In addition, a "light armored howitzer" and an "ammunition vehicle for the light armored howitzer" were planned. The intended production quotas were to be 45 and 6 units per month.

Beyond that, plans were made for an "Sturm Panzer III/IV" to be produced as of January 1945, at a rate of 20 units per month.

Finally, the "Flakpanzer III/IV Kugelblitz" was planned, which was to be produced at the rate of 30 units per month as of March 1945.

In the autumn of 1944 the idea of producing such a uniform chassis was finally dropped.

# The Chief Tank-Producing Firms

(as of 1943)

**Firms in the Special Branch of "Tank Production"**

Skoda AG, ex-Skodawerke, Königgrätz, Skoda
Bohemia Maschinenfabrik, Böhmisch-Leipa
Böhmisch-Mährische Maschinenfabrik, Prague X
Brandenburger Eisenwerke GmbH Brandenburg on the Havel, Werk Kirchmöser, Kirchmöser on the Havel
Gebr. Böhler & Co. AG, Kapfenberg-Deuchendoeutsche Edelstahlwerke AG, Hannover-Linden
Deutsche Edelstahlwerke AG, Krefeld
Deutsche Röhrenwerke, Werk Thyssen, Mülheim on the Ruhr
Dortmund-Hörder Hüttenverein, Werk Hörde, Dortmund-Hörde
Düsseldorfer Waggonfabrik, Düsseldorf
Eisenwerke Oberdonau GmbH, Linz on the Danube
Eisen-und Hüttenwerke AG, Bockum, Westphalia
Ferrum Werk AG, Laurahütte, Upper Silesia
Gutehoffnungshütte Oberhausen AG, Oberhausen, Rhineland
Harkort-Eicken Stahlwerke AG, Hagen, Westphalia
C.H. Jucho, Dortmund
Königs- und Bismarckhütte AG, Königshütte-Bismarckhütte, Upper Silesia
Werk Bankhütte, Dombrowa, Upper Silesia
Werk Hubertushütte, Hohenlinde, Upper Silesia
Friedrich Krupp AG, Essen on the Ruhr
Linke-Hofmann-Werke AG, Breslau
Heinrich Lanz AG, Mannheim
Press- und Ziehwerk GmbH, Krefeld-Urdingen
Ruhrstahl AG Heinrichshütte, Hettingen on the Ruhr
L. & C. Steinmüller, Gummersbach, Rhineland
Schöller Bleckmann Stahlwerke AG, Vienna
Witkowitzer Bergbau und Eisenhüttengewerkschaft, Mährisch-Ostrau
Poldihütte, Kladno

**Firms in the Special Branch of "Armored Combat Vehicle Production"**

Alkett, Berlin-Borsigwalde
Alkett, Berlin-Spandau
Böhmisch-Mährische Maschinenfabrik, Prague X
Borgward, Bremen
Daimler-Benz AG, Berlin-Marienfelde
Demag, Falkensee near Berlin, and Demag, Benrath
Deutsche Eisenwerke, Mülheim
Famo, Breslau 6
Henschel & Sohn GmbH, Kassel
Krupp-Gruson, Magdeburg
MAN Maschinenfabrik Augsburg-Nürnberg AG, Nürnberg 24
Miag, Braunschweig
MNH Maschinenfabrik Niedersachsen, Hannover-Wülfel
Nibelungen Werk, St. Valentin, Lower Danube
Stahlindustrie GmbH, Duisburg
Wegmann & Co., Kassel

**Firms in the Special Branch of "Armored Combat Vehicle Repair"**

Adlerwerke W. Seibert GmbH, AschaffenburgAG ex-Skodawerke Pilsen, Skoda
Daimler-Benz AG, Königsberg
FAMO Fahrzeug--und Motorenwerke, Breslau 6
Heeres-Kraftfahrzeug-Werkstatt, Vienna

Heeres-Kraftfahrzeug-Werkstatt, Pschelautsch, Bohemia
Klöckner-Humboldt-Deutz AG, Cologne-Deutz
Krupp-Druckenmüller GmbH, Berlin-Tempelhof
Kraus-Maffei AG, Munich-Allach
MAN Maschinenfabrik Augsburg-Nürnberg AG, Nürnberg 24
Nibelungenwerk, St. Valtin, Lower Danube
Richard Raupach, Warnsdorf, Sudeten District
Hermann Rüter, Langenhagen near Hannover

**Firms in the Special Branch "Special Motor Vehicles"**

Gustav Appel Maschinenfabrik, Berlin-Spandau
C.F.W. Borgward, Bremen
Büssing NAG, Leipzig N 26
Evens & Pistor, Helsa near Kassel
Rinco Motorenwerk, Rumburg, Sudeten District
Südeisenbau GmbH, Nürnberg 20
F. Schichau AG, Elbing, West Prussia
Weserhütte AG, Bad Oeynhausen, Westphalia
Wumag, Waggon- und Maschinenbau AG, Görlitz
A. Zachertz, Karosserie- und Fahrzeugfabrik, Freistadt, Lower Silesia
Zündapp-Werke GmbH, Nürnberg-West

**Firms in the Special Branch of "Tank Tracks"**

AG ex-Skodawerke Pilsen, Skoda
Ardeltwerke, Eberswalde in the Mark
Joachim Baumgart, Tönisheide, Post Velbert, Rhineland
Berg-und Hüttenwerke Ges. Karwin Trzynitz AG, Eisenwerk Trzynitz, Trzynitz, Upper Silesia
Bergisch-Mährisches Eisenwerk, Velbert, Rhineland
Bergische Stahlindustrie, Remscheid, Westphalia
Ed. Breitenbach GmbH, Weidenau on the Sieg
Brünn-Königsfelder Maschinen-und Waggonfabrik AG, Brünn-Königsfeld
Buderus'sche Eisenwerke, Wetzlar
Chiron-Werke GmbH, Tuttlingen, Württemberg
Demag AG, Wetter on the Ruhr
Deutsche Eisenwerke AG, Hilden/Ph.
von Duetrich & Co., Reichsofenwerk (Alsace)
Eisen-und Stahlwerke Hammerhütte, Ketschendorf on the Spree
Eisenwerke Gewerke R. und C. R. Lange, Hagen-Haspe
Elektro-Stahlwerk Wald GmbH, Solingen-Wald
August Engels GmbH, Velbert, Rhineland
August Engels GmbH, Delligsen Branch, Delligsen, Gandersheim District
Gusstahlwerk Wittmann AG, Hagen-Haspe
Gusstahlwerke Wolgast Kramer, Vogler & Co., Wolgast, Pomerania
Halbergerhütte GmbH, Brebach on the Saar
Albert Hoffmann, Eschweiler near Aachen
Max Jahn, Stahl- und Eisengeisserei GmbH, Leipzig W 35
Jenbacher Berg-und Hüttenwerke, Ernst Heinkel, Jenbach, Tirol
Otto Junker, Lammersdorf above Aachen
Knorr-Bremse AG, Volmarstein on the Ruhr
Krausewerk, Niederschl. Eisen-und Stahlwerke GmbH, Neusatz on the Oder
Krauss-Maffei AG, Munich-Allach
G. Krautheim AG, Chemnitz-Borna
Lindener Eisen-und Stahlwerke GmbH, Hannover-Linden
Mährische Stahl-und Eisenindustrie AG, Olmütz
Maschinen- und Werkzeugfabrik GmbH, Danzig-Neufahrwasser
Meier und Weichelt, Leipzig W 34
Metallgussges mbH, Leipzig W 35
Nationale Automobilges. AG, Abt. Prestowerke, Chemnitz
NSU-Werke AG, Neckarsulm
Oberhütten, Vereinigte Oberschles. Hüttenwerke AG, Abt. H 6, Gleiwitz
Ostschles. Eisenhüttenwerke "Osthütte" GmbH, Sosnowitz, Upper Silesia
Poldihütte, Komotau, Sudeten District
Reichsbahnausbesserungswerk Opladen, Motoren-Abteilung, Opladen
Rheinmetall-Borsig AG, Düsseldorf
Carl Ritscher GmbH, Hamburg-Harburg 1
Rondo-Werk, Schwelm, Westphalia
Ruhrstahl AG, Annener Gusstahlwerk, Witten-Annen on

the Ruhr
Saarländisches Stahlwerk Dingler, Karcher & Cie GmbH, Saarbrücken
Siemag, Siegener Maschinenbau AG Abt. Ingowerke, Eiserfeld on the Sieg
Siepmann-Werke, Belecke-Möhne
M. Schmid & Söhne, Wilhelmsburg, N.-D.
Schöller-Bleckmann Stahlwerke, Ternitz, N.-D.
Stahl-und Temperguss AG, ex-Fischer Traisen, Vienna IX
Stahlwerk Herbesthal GmbH, Herbesthal
M. Streicher. Stuttgart-Bad Canstatt
Vorwerk & Co., Wuppertal-Barmen
Wagner-Biro AG, Vienna V/35
Wichertler & Kowarik AG, Pressnitz, Moravia
MBA Maschinenbau und Bahnbedarf, Nordhausen


**Firms of the Special Branch "Motor Production"**
Adler-Werke AG, Frankfurt on the Main
Auto-Union, Siegmar, Saxony
Auto-Union, Zwickau
Borgward, Bremen
Henschel & Sohn GmbH, Kassel
Krauss-Maffei, Munich
MAN, Nürnberg
MBA Maschinenbau und Bahnbedarf, Nordhausen
Maybach Motorenbau GmbH, Friedrichshafen on the Bodensee
Nordbau, Berlin-Niederschöneweide
Saurer-Werke, Vienna
Simmering Graz Pauker AG, Vienna

**Firms of the Special Branch "Gear Production"**

Alkett, Berlin-Borsigwalde
Arbeitsgemeinschaft Getriebebau, Chemnitz
Arbeitsgemeinschaft Getriebebau, Heilbronn
Adler-Werke AG, Frankfurt on the Main
Heinrich Lanz AG, Mannheim
Steyr Daimler-Puch AG, Graz
Waldwerke GmbH, Passau
Zahnradfabrik Friedrichshafen AG, Friedrichshafen on the Bodensee
Zahnradfabrik Friedrichshafen AG, Berlin Branch, Berlin-Wittenau
Zahnradfabrik Friedrichshafen AG, Schwäbisch Gmünd Branch, Schwäbisch Gmünd

# Technical Data

## Panzerkampfwagen III (37mm) Ausf. A "1/ZW"

Manufacturer: Daimler-Benz AG, Berlin-Marienfelde Works
Years built: 1936-37

Source of Information: D 652/3 of January 25, 1938
Chassis numbers: 60 101-60 110

| | | | |
|---|---|---|---|
| Engine, make & type | Maybach HL 108 TR | Lubrication system | High-pressure central |
| Cylinders | 60-degree V-12 | Brake system | Mercedes-Benz mechanical |
| Bore & stroke | 110 x 115 mm | Brake type | Inner shoe |
| Displacement | 10,838 cc | Foot brake action | On driveshaft |
| Compression ratio | 6.5:1 | Hand brake action | On driveshaft |
| RPM, normal/maximum | 2600/3000 | Type of wheels | Road wheels and return rollers |
| Maximum power output | 230-250 HP | Track | 2047 mm |
| Power to weight | 17.3 HP/ton | Length of tracks | 3400 mm |
| Valve type | Dropped | Width of tracks | 360 mm |
| Crankshaft | 7 roller bearings | Ground clearance | 380 mm |
| Carburetors | 2 Solex 40 JFF II | Overall length | 5690 mm |
| Firing order | 1-8-5-10-3-7-6-11-2-9-4-12 | Overall width | 2810 mm |
| Starter | Bosch BNG 4/24 CRS 178 + AL/ZMD/R 10 | Overall height | 2335 mm |
| | | Ground pressure | 0.63 kp/sq. cm |
| Generator | Bosch GQL 300/12-900 RS 36 | Chassis weight | |
| Batteries | 2 12-volt 105-Ah | Allowable gross weight | 15,400 kp |
| Fuel pumps | 2 Solex pumps | Load limit | 1500 kp |
| Cooling | Water | Seats | 5 |
| Clutch | Dry multi-disc | Fuel consumption | 280 liters per 100 km |
| Gearbox | ZF SFG 75 shear | Oil consumption | Variable |
| Speeds | 5 forward, 1 reverse | Fuel supply | 300 liters |
| Drive wheels | Front | Armor: hull & turret | 14.5 mm all around |
| Drive axle ratio | | Performance: | |
| Top speed | 32 kph | Climbing ability | 30 degrees |
| Range | Road 165 km, off-road 95 km | Climbing distance | 600 mm |
| Steering | Daimler-Benz Winson clutch-type | Wading | 800 mm |
| Turning circle | | Spanning | 2600 mm |
| Suspension | Coil spring, 5-wheel running gear | Armament | 1 37mm KwK L/45 (120)* + 3 MG 34 (4425) |
| | | Intended use | Medium tank |

* Data in ( ) after armament are ammunition supplies

## Panzerkampfwagen III (37mm) Ausf. B "2/ZW"

Manufacturer: Daimler-Benz AG, Berlin-Marienfelde Works
Years built: 1937-38

Source of information: D 652/6 of February 7, 1938
Chassis numbers 60 201-60 215

| | | | |
|---|---|---|---|
| Engine, make & type | Maybach "HL 108 TR" | Lubrication system | High-pressure central |
| Cylinders | 60-degree V-12 | Brake system | Daimler-Benz mechanical |
| Bore x stroke | 110 x 115 mm | Brake type | Inner shoe |
| Displacement | 10,838 cc | Foot brake action | On driveshaft |
| Compression ratio | 6.5:1 | Hand brake action | On driveshaft |
| RPM, normal/maximum | 2600/3000 | Type of wheels | Road wheels and return rollers |
| Maximum power | 260/300 HP | Track | 2490 mm |
| Power to weight | 15.6 HP/ton | Track length | 3200 mm |
| Valve type | Dropped | Track width | 360 mm |
| Crankshaft | 7 roller bearings | Ground clearance | 375 mm |
| Carburetors | 2 Solex 40 JFF II | Overall length | 6000 mm |
| Firing order | 1-8-5-10-3-7-6-11-2-9-4-12 | Overall width | 2870 mm |
| Starter | Bosch BNG 4/24 CRS 178 + AL/ZMD/R 10 | Overall height | 2450 mm |
| | | Firing height | 1890 mm |
| Generator | Bosch GQL 300/12-900 RS 36 | Ground pressure | 0.65 kp/sq. cm |
| Batteries | 2 12-volt 105-Ah | Chassis weight | |
| Fuel pumps | 2 Solex pumps | Allowable gross weight | 15,900 kp |
| Cooling | Water | Load limit | 1500 kp |
| Clutch | Dry multi-disc | Seats | 5 |
| Gearbox | ZF SFG 75 shear | Fuel consumption | 280 liters/100 km |
| Speeds | 5 forward, 1 reverse | Oil consumption | Variable |
| Drive wheels | Front | Fuel capacity | 300 liters |
| Drive axle ratio | | Armor: hull & turret | 14.5 mm all around |
| Top speed | 35 kph | Performance: | |
| Range | road 165, off-road 95 km | Climbing ability | 30 degrees |
| Steering | Daimler-Benz/Wilson clutch-type | Climbing distance | 600 mm |
| Turning circle | 5.85 meters | Wading | 800 mm |
| Suspension | 2 leaf springs on each side, each for one group of 4 road wheels | Spanning | 2300 mm |
| | | Armament | 1 37mm KwK L/45 (120) + 3 MG 34 (4425) |
| | | Intended use | Medium tank |

# Panzerkampfwagen III (37mm) Ausf. C "3a/ZW"

Manufacturer: Daimler-Benz AG, Berlin-Marienfelde Works
Years built: 1937-38

Source of information: D 652/6 of February 7, 1938
Chassis numbers: 60 301-60 315

| | | | |
|---|---|---|---|
| Engine, make & type | Maybach "HL 108 TR" | Lubrication system | High-pressure central |
| Cylinders | 60-degree V-12 | Brake system | Daimler-Benz mechanical |
| Bore x stroke | 100 x 115 mm | Brake type | Inner shoe |
| Displacement | 10,838 cc | Foot brake action | On driveshaft |
| Compression ratio | 6.5:1 | Hand brake action | On driveshaft |
| RPM, normal/maximum | 2600/3000 | Wheels | Road wheels & return rollers |
| Maximum power | 230-250 HP | Track | 2490 mm |
| Power to weight | 15.6 HP/ton | Track length | 3200 mm |
| Valves | dropped | Track width | 360 mm |
| Crankshaft | 7 roller bearings | Ground clearance | 375 mm |
| Carburetors | 2 Solex 40 JFF II | Overall length | 5690 mm |
| Firing order | 1-8-5-10-3-7-6-11-2-9-4-12 | Overall width | 2810 mm |
| Starter | Bosch BNG 4/24 CRS 178 + AL/ZMD/R 10 | Overall height | 2335 mm |
| | | Firing height | 1890 mm |
| Generator | Bosch GQL 300/12-900 RS 36 | Ground pressure | 0.65 kp/sq. cm |
| Batteries | 2 12-volt 105-Ah | Chassis weight | |
| Fuel pumps | 2 Solex pumps | Allowable gross weight | 15,900 kp |
| Cooling | Water | Load limit | 1500 kp |
| Clutch | Dry multi-plate | Seats | 5 |
| Gearbox | ZF SFG 75 shear | Fuel consumption | 280 liters/100 km |
| Speeds | 5 forward, 1 reverse | Oil consumption | Variable |
| Drive wheels | Front | Fuel capacity | 300 liters |
| Drive axle ratio | | Armor, hull & turret | 14.5 mm all around |
| Top speed | 35 kph | Performance: | |
| Range | road 165, off-road 95 km | Climbing ability | 30 degrees |
| Steering | Daimler-Benz/Wilson clutch-type | Climbing distance | 600 mm |
| Turning circle | 5.85 meters | Wading | 800 mm |
| Suspension | 1 leaf spring each for 2 front, 2 rear, & 4 central road wheels | Spanning | 2300 mm |
| | | Armament | 1 37mm KwK L/45 (120) + 3 MG 34 (4425) |
| | | Intended use | Medium tank |

# Panzerkampfwagen III (37mm) (Sd.Kfz.141) Ausf. D "3b/ZW"

Manufacturer: Daimler-Benz AG, Berlin-Marienfelde Works
Year built: 1938

Source of information: D 652/12 of December 15, 1938
Chassis numbers: 60 221-60 225 and 60 316-60 340

| | | | |
|---|---|---|---|
| Engine make & type | Maybach "HL 108 TR" | Lubricating system | High-pressure central |
| Cylinders | 60-degree V-12 | Brake system | Daimler-Benz mechanical |
| Bore x stroke | 100 x 115 mm | Brake type | Inner shoe |
| Displacement | 10,838 cc | Foot brake action | On driveshaft |
| Compression ratio | 6.5:1 | Hand brake action | On driveshaft |
| RPM normal/maximum | 2600/3000 | Wheels | Road wheels & return rollers |
| Maximum power | 230-250 HP | Track | 2490 mm |
| Power to weight | 12.5 HP/ton | Track length | 3200 mm |
| Valves | dropped | Track width | 360 mm |
| Crankshaft | 7 roller bearings | Ground clearance | 375 mm |
| Carburetors | 2 Solex 40 JFF II | Overall length | 6000 mm |
| Firing order | 1-8-5-10-3-7-6-11-2-9-4-12 | Overall width | 2870 mm |
| Starter | Bosch BNG 4/24 CRS 178 + AL/ZMD/R 10 | Overall height | 2450 mm |
| | | Firing height | 1890 mm |
| Generator | Bosch GQL 300/12-900 RS 36 | Ground pressure | 0.95 kp/sq. cm |
| Batteries | 2 12-volt, 105-Ah | Chassis weight | 13,300 kp |
| Fuel pumps | 2 Solex pumps | Allowable gross weight | 19,800 kp |
| Cooling | Water | Load limit | 1500 kp |
| Gearbox | ZF SSG 76 shear | Seats | 5 |
| Speeds | 6 forward, 1 reverse | Fuel consumption | 280 liters/100 km |
| Drive wheels | Front | Oil consumption | Variable |
| Drive axle ratio | | Fuel capacity | 300 liters |
| Top speed | 35 kph | Armor: hull & turret | 30 mm all around |
| Range | Road 165, off-road 95 km | Performance: | |
| Steering | Daimler-Benz/Wilson clutch-type | Climbing ability | 30 degrees |
| Turning circle | 5.85 meters | Climbing distance | 600 mm |
| Suspension | 3 leaf springs per side, but modified system | Wading | 800 mm |
| | | Spanning | 2300 mm |
| | | Armament | 1 37mm KwK L/45 (120) + 3 MG 34 (4425) |
| | | Intended use | Medium tank |

**Panzerkampfwagen III (37mm)**
**(Sd.Kfz.141) Ausf. E "4/ZW"**

Daimler-Benz factory designation: "ZW 38"
Manufacturer: Daimler-Benz AG, Berlin-Marienfelde and other works

Year built: 1939
Source of information: D 652/17 of April 23, 1940
Chassis numbers: 60 401-60 441 and between 60 401 and 61 000 (without 60 501-60 545)

| | | | |
|---|---|---|---|
| Engine: make & type | Maybach "HL 120 TR" | Brake action | Hydraulic with servo |
| Cylinders | 60-degree V-12 | Brake type | Inner shoe |
| Bore x stroke | 105 x 115 mm | Foot brake action | On drive wheels |
| Displacement | 11,867 cc | Hand brake action | On drive wheels |
| Compression ratio | 6.2-6.5:1 | Wheels | Road wheels 520 x 95-398, |
| RPM normal/maximum | 2600/3000 | | Return rollers 310 x 70-203 |
| Maximum power | 265-300 HP | Track | 2490 mm |
| Power to weight | 14.8 HP/ton | Track length | 2860 mm |
| Valves | Dropped | Track width | 360 mm |
| Crankshaft | 7 roller bearings | Ground clearance | 385 mm |
| Carburetors | 2 Solex 40 JFF II | Overall length | 5380 mm |
| Firing order | 1-12-5-8-3-10-6-7-2-11-4-9 | Overall width | 2910 mm |
| Starter | Bosch BNG 4/24 ARS 129 + AL/ZMD/R 9 | Overall height | 2435 mm |
| | | Ground pressure | 0.99 kp/sq. cm |
| Maximum torque | 80 m/kg | Chassis weight | 13,800 kp |
| Engine weight | 920 kp | Allowable gross weight | 19,500 kp |
| Generator | Bosch GTLN 700/12-1500 | Load limit | 1500 kp |
| Batteries | 2 12-volt, 105-Ah | Seats | 5 |
| Fuel pumps | 2 mechanical pumps | Fuel consumption | 187 liters/100 km or 235-255 g/HP/h |
| Cooling | Water | | |
| Clutch | Hydraulic, with accelerator | Oil consumption | Variable |
| Gearbox | Maybach SGR 32 8 145-Variorex (pre-selector) | Fuel capacity | 320 liters |
| | | Armor: hull & turret | 30 mm all around |
| Speeds | 10 forward, 1 reverse | Performance: | |
| Drive wheels | Front | Climbing ability | 30 degrees |
| Drive axle ratio | 1:4 (side rods) | Climbing distance | 600 mm |
| Top speed | 40 kph | Wading | 800 mm |
| Range | Road 165, off-road 95 km | Spanning | 2300 mm |
| Steering | Daimler-Benz/Wilson clutch-type | Armament | 1 37mm KwK L/45 (120) + 2 MG 34 (3600) |
| Turning circle | 5.85 meters | | |
| Suspension | 12 transverse torsion bars | Intended use | Medium tank, later rearmed with 50mm KwK L/42 |
| Lubricating system | High-pressure | | |
| Brake system | Daimler-Benz | | |

## Panzerkampfwagen III (37mm) (Sd.Kfz.141) Ausf. F "5/ZW"

Daimler-Benz factory designation: "ZW 38"
Manufacturer: Daimler-Benz AG, Berlin-Marienfelde and other works

Year built: 1939
Source of information: D 652/17 of April 23, 1940
Chassis numbers: 61 001-65 000

| | | | |
|---|---|---|---|
| Engine make & type | Maybach "HL 120 TRM" | Brake type | Inner shoe |
| Cylinders | 60-degree V-12 | Foot brake action | On drive wheels |
| Bore x stroke | 105 x 115 mm | Hand brake action | On drive wheels |
| Displacement | 11,867 cc | Wheel type | Road wheels 520 x 95-398 |
| Compression ratio | 6.2-6.5:1 | | Return rollers 310 x 70-203 |
| RPM normal/maximum | 2600/3000 | Track | 2490 mm |
| Maximum power | 265-300 HP | Track length | 2860 mm |
| Power to weight | 14.8 HP/ton | Track width | 360 mm |
| Valves | Dropped | Ground clearance | 385 mm |
| Crankshaft | 7 roller bearings | Overall length | 5380 mm |
| Carburetors | 2 Solex 40 JFF II | Overall width | 2910 mm |
| Firing order | 1-12-5-8-3-10-6-7-2-11-4-9 | Overall height | 2435 mm |
| Starter | Bosch BNG 4/24 ARS 129 + Bosch AL/ZMD/R 9 | Ground pressure | 0.99 kp/sq. cm |
| | | Chassis weight | 13,800 kp |
| Generator | Bosch GTLN 700/12-1500 | Allowable gross weight | 19,500 kp |
| Batteries | 2 12-volt 1-5-Ah | Load limit | 1500 kp |
| Fuel pumps | 2 pumps | Seats | 5 |
| Cooling | Water | Fuel consumption | 187 liters/100 km |
| Clutch | Hydraulic, with accelerator | Oil consumption | Variable |
| Gearbox | Maybach "SRG 328 145"-Variorex | Fuel capacity | 320 liters |
| | | Armor, hull and turret | 30 mm all around |
| Speeds | 10 forward, 1 reverse | Performance: | |
| Drive wheels | Front | Climbing ability | 30 degrees |
| Drive axle ratio | 1:4 (side rods) | Climbing distance | 600 mm |
| Top speed | 40 kph | Wading | 800 mm |
| Range | road 165, off-road 95 km | Spanning | 2300 mm |
| Steering | Daimler-Benz/Wilson clutch-type | Armament | 1 37mm KwK L/45 (120), later 50mm KwK 39 L/42 (99); 2 MG 34 (3750) |
| Turning circle | 5.85 meters | | |
| Suspension | 12 transverse torsion bars | | |
| Lubrication system | High-pressure | Intended use | Medium tank |
| Brake system | Daimler-Benz hydraulic, with servo | | |

## Panzerkampfwagen III (37mm) (Sd.Kfz.141) Ausf. G "6/ZW"

DB factory designation "ZW 38"

Manufacturer: Daimler-Benz AG, Berlin-Marienfelde and other works

Years built: 1939-1940
Source of information: D 652/19 of March 25, 1942
Chassis numbers 65 001-66 000

| | | | |
|---|---|---|---|
| Engine make & type | Maybach "HL 120 TRM" | Brake type | Inside shoe |
| Cylinders | 60-degree V-12 | Foot brake action | On drive wheels |
| Bore x stroke | 105 x 115 mm | Hand brake action | On drive wheels |
| Displacement | 11,867 cc | Wheel type | Road wheels 520 x 95-398 |
| Compression ratio | 6.2-6.5:1 | | Return rollers 310 x 70-203 |
| RPM normal/maximum | 2600/3000 | Track | 2490 mm |
| Maximum power | 265/300 HP | Track length | 2860 mm |
| Power to weight | 14.8 HP/ton | Track width | 360 mm |
| Valves | Dropped | Ground clearance | 385 mm |
| Crankshaft | 7 roller bearings | Overall length | 5380 mm |
| Carburetors | 2 Solex 40 JFF II | Overall width | 2910 mm |
| Firing order | 1-12-5-8-3-10-6-7-2-11-4-9 | Overall height | 2435 mm |
| Starter | Bosch BNG 4/24 ARS 129 + AL/ZMD/R 9 | Ground pressure | 0.99 kp/sq. cm |
| | | Chassis weight | 13,800 kp |
| Generator | Bosch GQL 300/12-900 | Allowable gross weight | 19,500 kp |
| Batteries | 2 12-volt 105-Ah | Load limit | 1500 kp |
| Fuel pumps | 2 mechanical, 1 electric | Seats | 5 |
| Cooling | Water | Fuel consumption | 235-255 g/HP/h; 187 liters/100km |
| Clutch | Hydraulic with pre-selector | Oil consumption | Variable |
| Gearbox | Maybach SRG 328145 Variorex | Fuel capacity | 320 liters |
| Speeds | 10 forward, 1 reverse | Armor: hull and turret | 30 mm all around |
| Drive wheels | Front | Performance: | |
| Drive axle ratio | 1:4 (side rods) | Climbing ability | 30 degrees |
| Top speed | 40 kph | Climbing distance | 600 mm |
| Range | Road 165, off-road 95 km | Wading | 800 mm |
| Steering | DB/Wilson clutch-type, hydraulically supported | Spanning | 2300 mm |
| | | Armament | 1 37mm KwK L/45 (120), later 50mm KwK 39 L/42 (99), + 2 MG 34 (3750) |
| Turning circle | 5.85 meters | | |
| Suspension | 12 transverse torsion bars, | | |
| Lubrication | High-pressure | Intended use | Medium tank |
| Brake system | Daimler-Benz hydraulic, with servo | | |

## Panzerkampfwagen III (50mm) (Sd.Kfz.141), Ausf. H "7/ZW"

Made in Germany
Manufacturer: Daimler-Benz AG, Berlin-Marienfelde and other works

Years built: 1940-1941
Source of information: D 652/26 of November 1, 1942
Chassis numbers: 66 001-68 000

| | | | |
|---|---|---|---|
| Engine make & type | Maybach "HL 120 TRM" | Brake system | Daimler-Benz hydraulic with servo |
| Cylinders | 60-degree V-12 | | |
| Bore x stroke | 105 x 115 mm | Brake type | Inner shoe |
| Displacement | 11,867 cc | Foot brake action | On drive wheels |
| Compression ratio | 6.2-6.5:1 | Hand brake action | On drive wheels |
| RPM normal/maximum | 2600/3000 | Wheel type | Road wheels 520/95-397 |
| Maximum power | 265/300 HP | | Return rollers 310/70-203 |
| Power to weight | 13.7 HP/ton | Track | 2510 mm |
| Valves | Dropped | Track length | 2860 mm, type Kgs 61/400/120 |
| Crankshaft | 7 roller bearings | Track width | 400 mm, 93 links |
| Carburetors | 2 Solex 40 JFF II | Ground clearance | 385 mm |
| Firing order | 1-12-5-8-3-10-6-7-2-11-4-9 | Overall length | 5520 mm |
| Starter | Bosch BNG 4/24 ARS 129 + AL/ZMD/R 9 | Overall width | 2950 mm |
| | | Overall height | 2500 mm |
| Generator | Bosch GTLN 600/12-1500 | Ground pressure | 0.93 kp/sq. cm |
| Batteries | 2 12-volt 105-Ah | Chassis weight | 15,800 kp |
| Fuel pumps | 2 mechanical & 1 electric | Allowable gross weight | 21,600 kp |
| Cooling | Water | Load limit | 1500 kp |
| Clutch | F & S dry 3-plate La 120 HDA | Seats | 5 |
| Gearbox | ZF SSG 77 Aphon | Fuel consumption | 182 liters/100 km |
| Gears | 6 forward, 1 reverse | Oil consumption | Variable |
| Drive wheels | Front | Fuel capacity | 320 liters |
| Drive axle ratio | 1:4 (side rods) | Armor: hull & turret | 30 mm all around |
| Top speed | 40 kph | Performance: | |
| Range | Road 165, off-road 95 km | Climbing ability | 30 degrees |
| Steering | Daimler-Benz/Wilson clutch-type, hydraulically supported | Climbing distance | 600 mm |
| | | Wading | 800 mm |
| Turning circle | 5.85 meters | Spanning | 2590 mm |
| Suspension | 12 transverse torsion bars | Armament | 1 50mm KwK 39 L/42 (99) + 2 MG 34 (3750) |
| Lubrication | High-pressure | | |
| | | Intended use | Medium tank |

## Panzerkampfwagen III (50mm) (Sd.Kfz.141) Ausf. J "8/ZW"

Manufacturer: Daimler-Benz AG, Berlin-Marienfelde and other works
Year built: 1941

Source of information: D 652/26 of November 1, 1942
Chassis numbers: 68 001-69 100 & 72 001-74 100
Data in ( ) applies to 72 001-74 000 version.

| | | | |
|---|---|---|---|
| Engine make & type | Maybach "HL 120 TRM", TypeA | Hand brake action | On drive wheels |
| Cylinders | 60-degree V-12 | Wheel type | Road wheels 520 x 95-397 |
| Bore x stroke | 105 x 115 mm | | Return rollers 310 x 70-203 |
| Displacement | 11,867 cc | Track | 2510 mm |
| Compression ratio | 6.2-6.5:1 | Track type | Kgs 61/400/120 |
| RPM normal/maximum | 2600/3000 | Track length | 2860 mm, 93 links |
| Maximum power | 265/300 HP | Track width | 400 mm |
| Power to weight | 13.5 HP/ton | Ground clearance | 385 mm |
| Valves | Dropped | Overall length | 5560 mm |
| Crankshaft | 7 roller bearings | Overall width | 2950 mm |
| Carburetors | 2 Solex 40 JFF II | Overall height | 2500 mm |
| Firing order | 1-12-5-8-3-10-6-7-2-11-4-9 | Firing height | 1900 mm |
| Starter | Bosch BNG 4/24 ARS 129 + Bosch AL/ZMD/R 9 | Ground pressure | 0.94 kp/sq. cm |
| | | Chassis weight | 15,800 kp |
| Generator | Bosch GTLN 600/12-1500 | Allowable gross weight | 21,500 kp |
| Batteries | 2 12-volt 125-Ah | Load limit | 1500 kp |
| Fuel pumps | 2 mechanical & 1 electric | Seats | 5 |
| Cooling | Water | Fuel consumption | 182 liters/100 km |
| Clutch | F & S dry 3-plate La 120 HDA | Oil consumption | Variable |
| Gearbox | ZF SSG 77 Aphon | Fuel capacity | 320 liters |
| Gears | 6 forward, 1 reverse | Armor: | |
| Drive wheels | Front | Hull front | 50 mm |
| Drive axle ratio | 1:4 (side rods) | Hull sides | 30 mm |
| Top speed | 40 kph | Hull rear | 50 mm |
| Range | Road 145, off-road 85 km | Turret front | 50 mm |
| Steering | Daimler-Benz/Wilson clutch-type | Turret sides & rear | 30 mm |
| Turning circle | 5.85 meters | Performance: | |
| Suspension | 12 transverse torsion bars | Climbing ability | 30 degrees |
| Lubrication | High-pressure | Climbing distance | 600 mm |
| Brake system | Daimler-Benz mechanical | Wading | 800 mm |
| Brake type | Inner shoe | Spanning | 200 mm |
| Foot brake action | On drive wheels | Armament | 1 50mm KwK 39 L/42 (L/60 as of chassis 72 001) (99/84) + 2 MG 34 (3750) |
| | | Intended use | Medium tank |

# Panzerkampfwagen III (50mm) (Sd.Kfz.141/1) Ausf. L "9/ZW"

Daimler-Benz factory designation "ZW 38"

Manufacturer: Daimler-Benz AG, Berlin-Marienfelde and other works

Years built: 1941-42
Source of information: D 652/26 of November 1, 1942
Chassis numbers 74 101-76 000

| | | | |
|---|---|---|---|
| Engine make & type | Maybach "HL 120 TRM" | Foot brake action | On drive wheels |
| Cylinders | 60-degree V-12 | Hand brake action | On drive wheels |
| Bore x stroke | 105 x 115 mm | Wheel type | Road wheels 520/95-398 |
| Displacement | 11,867 cc | | Return rollers 310/70-203 |
| Compression ratio | 6.2-6.5:1 | Track | 2510 mm |
| RPM normal/maximum | 2600/3000 | Track length | 2860 mm – type Kgs 61/400/120 |
| Maximum power | 265/300 HP | Track width | 400 mm – 93 links |
| Power to weight | 14.0 HP/ton | Ground clearance | 385 mm |
| Valves | dropped | Overall length | 5560 minus gun, 6412 in all |
| Crankshaft | 7 roller bearings | Overall width | 2950 mm |
| Carburetors | 2 Solex 40 JFF II | Overall height | 2500 mm |
| Firing order | 1-12-5-8-3-10-6-7-2-11-4-9 | Firing height | 1900 mm |
| Starter | Bosch BNG 4/24 ARS 129 + AL/ZMD/R 9 | Ground pressure | 0.94 kp/sq. cm |
| | | Chassis weight | 15,800 kp |
| Generator | Bosch GTLN 600/12-1500 | Allowable gross weight | 21,300 kp |
| Batteries | 1 12-volt a5-Ah | Load limit | 1500 kp |
| Fuel pumps | 2 Solex mechanical + 1 electric | Seats | 5 |
| Cooling | Water | Fuel consumption | 182 liters/100 km |
| Clutch | F & S dry 3-plate La 120/HDA | Oil consumption | Variable |
| Gearbox | ZF SSG 77 Aphon | Fuel capacity | 320 liters |
| Speeds | 6 forward, 1 reverse | Armor: hull front | 50 + 20 mm |
| Drive wheels | Front | sides | 30 mm |
| Drive axle ratio | 1:4 (side rods) | rear | 50 mm |
| Top speed | 40 kph | turret front | 50 + 20 mm |
| Range | Road 155, off-road 95 km | sides + rear | 30 mm |
| Steering | Daimler-Benz/Wilson clutch-type | Performance: | |
| Turning circle | 5.85 meters | Climbing ability | 30 degrees |
| Suspension | 12 transverse torsion bars | Climbing distance | 600 mm |
| Lubrication | High-pressure | Wading | 800 mm |
| Brake system | Daimler-Benz mechanical | Spanning | 2000 mm |
| Brake type | Inner shoe | Armament | 1 50mm KwK 39 L/60 (78) + 2 MG 34 (4950) |
| | | Intended use | Medium tank |

# Panzerkampfwagen III (50mm) (Sd.Kfz.141/1) Ausf. M "10/ZW"

Daimler-Benz factory designation "ZW 38"

Manufacturer: Daimler-Benz AG, Berlin-Marienfelde and other works

Year built: 1942
Source of information: D 652/26 of November 1, 1942
Chassis numbers 76 001-78 000

| | | | |
|---|---|---|---|
| Engine make & type | Maybach "HL 120 TRM" | Wheel type | Road wheels 520/95-398 |
| Cylinders | 60-degree V-12 | | Return rollers 310/70-203 |
| Bore x stroke | 105 x 115 mm | Track | 2510 mm |
| Displacement | 11,867 cc | Track length | 860 mm, type Kgs 61/400/120 |
| Compression ratio | 6.2-6.5:1 | Track width | 400 mm, 93 links |
| RPM normal/maximum | 2600/3000 | Ground clearance | 385 mm |
| Maximum power | 265/300 HP | Overall length | 5560 mm minus gun, 6412 mm in all |
| Power to weight | 14.2 HP/ton | | |
| Valves | Dropped | Overall width | 970 mm, with aprons 3266 mm, with eastern tracks 3410 mm |
| Crankshaft | 7 roller bearings | | |
| Carburetors | 2 Solex 40 JFF II | Overall height | 2500 mm |
| Firing order | 1-12-5-8-3-10-6-7-2-11-4-9 | Firing height | 1900 mm |
| Starter | Bosch BNG 4/24 ARS 129 + AL/ZMD/R 9 | Ground pressure | 0.94 kp/sq. cm |
| | | Chassis weight | 15,800 kp |
| Generator | Bosch GTLN 600/12-1500 | Allowable gross weight | 21,130 kp |
| Batteries | 2 12-volt 105-Ah | Load limit | 1500 kp |
| Fuel pumps | 2 Solex mechanical + 1 electric | Seats | 5 |
| Cooling | Water | Fuel consumption | 182 liters/100 km |
| Clutch | F & S dry 3-plate La 120/HDA | Oil consumption | Variable |
| Gearbox | ZF SSG 77 Aphon | Fuel capacity | 320 liters |
| Speeds | 6 forward, 1 reverse | Armor: hull front | 50 + 20 mm |
| Drive wheels | Front | sides | 30 mm |
| Drive axle ratio | 1:4 (side rods) | rear | 50 mm |
| Tope speed | 40 kph | turret front | 50 mm |
| Range | Road 155, off-road 95 km | sides + rear | 30 mm |
| Steering | Daimler-Benz/Wilson clutch-type | Performance: | |
| Turning circle | 5.85 meters | Climbing ability | 30 degrees |
| Suspension | 12 transverse torsion bars | Climbing distance | 600 mm |
| Lubrication | High-pressure | Wading | 800 mm |
| Brake system | Daimler-Benz mechanical | Spanning | 2000 mm |
| Brake type | Inner shoe | Armament | 1 50mm KwK 39 L/60 (84) or 75mm KwK L/24, + 2 MG 34 (3800) |
| Foot brake action | On drive wheels | | |
| Hand brake action | On drive wheels | | |
| | | Intended use | Medium tank |

## Panzerkampfwagen III (Fl) (Sd.Kfz.141/3) "10/ZW"

Manufacturer: Mühlenbau- und Industrie AG (MIAG), Amme works, Braunschweig
Flamethrower installed by: Waggonfabrik Wegmann AG, Kassel

Year built: 1942
Source of information: WaA Handbook, Section G 306
Note: 100 vehicles were made.

| | | | |
|---|---|---|---|
| Engine make & type | Maybach "HL 120 TRM" | Hand brake action | On drive wheels |
| Cylinders | 60-degree V-12 | Wheel type | Road wheels 520 x 95-398 |
| Bore x stroke | 105 x 115 mm | | Return rollers 310 x 70-203 |
| Displacement | 11,867 cc | Track | 2510 mm |
| Compression ratio | 6.2-6.5:1 | Track length | 2869 mm, 93 links |
| RPM normal/maximum | 2600/3000 | Track width | 400 mm |
| Maximum power | 265/300 HP | Ground clearance | 385 mm |
| Power to weight | 13.0 HP/ton | Overall length | 6408 mm |
| Valves | Dropped | Overall width | 2970/3266 mm with eastern tracks, 3410 with aprons |
| Crankshaft | 7 roller bearings | | |
| Carburetors | 2 Solex 40 JFF II | Overall height | 2500 mm |
| Firing order | 1-12-5-8-3-10-6-7-2-11-4-9 | Ground pressure | 1.03 kp/sq. cm |
| Starter | Bosch BNG 4/24 ARS 129 + Bosch AL/ZMD/R 9 | Chassis weight | 15,800 kp |
| | | Allowable gross weight | 23,000 kp |
| Generator | Bosch GTLN 600/12-1500 | Load limit | 1500 kp |
| Batteries | 2 12-volt 105-Ah | Seats | 3 |
| Fuel pumps | 2 mechanical + 1 electric | Fuel consumption | 182 liters/100 km |
| Cooling | Water | Oil consumption | Variable |
| Clutch | F & S dry 3-plate La 120/HDA | Armor: hull front | 50 + 30 mm |
| Gearbox | ZF SSG 77 Aphon | sides | 30 mm |
| Speeds | 6 forward, 1 reverse | rear | 50 mm |
| Drive wheels | Front | turret front | 50 + 20 mm |
| Drive axle ratio | 1:4 (side rods) | side & rear | 30 mm |
| Top speed | 40 kph | Performance: | |
| Range | road 155, off-road 95 km | Climbing ability | 30 degrees |
| Steering | Daimler-Benz/Wilson clutch-type | Climbing distance | 600 mm |
| Turning circle | 5.85 meters | Wading | 800 mm |
| Track type | Kgs 61/400/120 | Spanning | 2000 mm |
| Suspension | 12 transverse torsion bars | Armament | 1 flamethrower (1000 liters oil, 14 mm jet, electric ignition, range 55-60 m, + 2 MG 34 |
| Lubrication | High-pressure | | |
| Brake system | Daimler-Benz mechanical | | |
| Brake type | Inner shoe | Intended use | Flamethrowing tank |
| Foot brake action | On drive wheels | | |

## Panzerkampfwagen III (75mm) (Sd.Kfz.141/2) Ausf. N "11/ZW"

Manufacturer: Daimler-Benz AG, Berlin-Marienfelde and other works
Year built: 1943

Source of information: D 652/26 of November 1, 1942
Chassis numbers start at 78 001
Final version of Panzer III

| | | | |
|---|---|---|---|
| Engine make & type | Maybach "HL 120 TRM" | Foot brake action | On drive wheels |
| Cylinders | 60-degree V-12 | Hand brake action | On drive wheels |
| Bore x stroke | 105 x 115 mm | Wheel type | Road wheels 520 x 95-398 |
| Displacement | 11,867 cc | | Return rollers 310 x 70-203 |
| Compression ratio | 6.2-6.5:1 | Track | 2510 mm |
| RPM normal/maximum | 2600/3000 | Track length | 2860 mm, 93 links |
| Maximum power | 265/300 HP | Track width | 400 mm |
| Power to weight | 13.0 HP/ton | Ground clearance | 385 mm |
| Valves | Dropped | Overall length | 5650 mm |
| Crankshaft | 7 roller bearings | Overall width | 2970/3266 mm with eastern tracks, |
| Carburetors | 2 Solex 40 JFF II | | 3410 mm with aprons |
| Firing order | 1-12-5-8-3-10-6-7-2-11-4-9 | | |
| Starter | Bosch BNG 4/24 ARS 129 + Bosch AL/ZMD/R 9 | Overall height | 2500 mm |
| | | Firing height | 1900 mm |
| Generator | Bosch GTLN 600/12-1500 | Ground pressure | 1.03 kp/sq. cm |
| Batteries | 2 12-volt 105-Ah | Chassis weight | 15,800 kp |
| Fuel pumps | 2 mechanical + 1 electric | Allowable gross weight | 21,300/23 000 kp |
| Cooling | Water | Load limit | 1700 kp |
| Clutch | F & S dry 3-plate La 120/HDA | Seats | 5 |
| Gearbox | ZF SSG 77 Aphon | Fuel consumption | 182 liters/100 km |
| Speeds | 6 forward, 1 reverse | Oil consumption | Variable |
| Drive wheels | Front | Fuel capacity | 320 liters |
| Drive axle ratio | 1:4 (side rods) | Armor: hull front | 50 + 20 mm |
| Top speed | 40 kph | sides | 30 mm |
| Range | road 155, off-road 95 km | rear | 50 mm |
| Steering | Daimler-Benz/Winson clutch-type | turret front | 50 mm |
| Turning circle | 5.85 meters | sides & rear | 30 mm |
| Track type | Kgs 61/400/120 | Performance: | |
| Suspension | 12 transverse torsion bars | Climbing ability | 30 degrees |
| Lubrication | High-pressure | Climbing distance | 600 mm |
| Brake system | Daimler-Benz mechanical | Wading | 800 mm |
| Brake type | Inner shoe | Spanning | 2000 mm |
| | | Armament | 1 75mm KwK L/24 (64) + 2 MG 34 (3450) |

## Gerpanzerte Selbstfahrlafette für Sturmgeschütz 75mm Sturmkanone (Sd.Kfz.142) Ausf. A "5/ZW"

Manufacturer: Altmärkische Kettenwerk GmbH, Berlin-Spandau  
Year built: 1940

Source of information: D 652 of April 1, 1943  
Chassis numbers 90 001-90 100

| | | | |
|---|---|---|---|
| Engine make & type | Maybach "HL 120 TR" | Brake type | Inner shoe |
| Cylinders | 60-degree V-12 | Foot brake action | On drive wheels |
| Bore x stroke | 105 x 115 mm | Hand brake action | On drive wheels |
| Displacement | 11,867 cc | Wheel type | Road wheels 520 x 95-398 |
| Compression ratio | 6.2-6.5:1 | | Return rollers 310 x 70-203 |
| RPM normal/maximum | 2600/3000 | Track | 2490 mm |
| Maximum power | 265/300 HP | Track length | 2860 mm |
| Power to weight | 15.3 HP/ton | Track width | 360 mm |
| Valves | dropped | Ground clearance | 385 mm |
| Crankshaft | 7 roller bearings | Overall length | 5380 mm |
| Carburetors | 2 Solex 40 JFF II | Overall width | 2920 mm |
| Firing order | 1-12-5-8-3-10-6-7-2-11-4-9 | Overall height | 1950 mm |
| Starter | Bosch BNG 4/24 ARS 129 + AL/ZMD/R 9 | Firing height | 1500 mm |
| | | Ground pressure | 0.093 kp/sq. cm |
| Generator | Bosch GTLN 700/12-1500 | Chassis weight | 13,800 kp |
| Batteries | 2 12-volt 105-Ah | Allowable gross weight | 19,600 kp |
| Fuel pumps | 2 Solex + 1 electric | Load limit | 1500 kp |
| Cooling | Water | Seats | 4 |
| Clutch | Hydraulic with accelerator | Fuel consumption | 187 liters/100 km |
| Gearbox | Maybach SRG 32 8 145-Variorex | Oil consumption | Variable |
| Speeds | 10 forward, 1 reverse | Fuel capacity | 310 liters |
| Drive wheels | Front | Armor: hull front | 50 mm |
| Drive axle ratio | 1:4 (side rods) | sides & rear | 30 mm |
| Top speed | 30 kph | Performance: | |
| Range | Road 160, off-road 100 km | Climbing ability | 30 degrees |
| Steering | Daimler-Benz/Wilson clutch-type | Climbing distance | 600 mm |
| Turning circle | 5.85 meters | Wading | 1000 mm |
| Suspension | 12 transverse torsion bars | Spanning | 2300 mm |
| Lubrication | High-pressure | Armament | 1 75mm StuK L/24 (44) + 2 MP |
| Brake system | Daimler-Benz hydraulic, with servo | Intended use | Armored escort for infantry + antitank forces |

## Gerpanzerte Selbstfahrlafette für Sturmgeschütz 75mm Sturmkanone (Sd.Kfz.142) Ausf. B-D "7/ZW"

Manufacturer: Altmärkische Kettenfabrik GmbH, Berlin-Spandau
Years built: 1940-41

Source of information: D 652/41 of April 1, 1943
Chassis numbers: Ausf. B: 90 101-90 550; Ausf. C: 90 551-90 600, Ausf. D: 90 601-90 750

| | | | |
|---|---|---|---|
| Engine make & type | Maybach "HL 120 TRM" | Brake system | Daimler-Benz hydraulic, with servo |
| Cylinders | 60-degree V-12 | | |
| Bore x stroke | 105 x 115 mm | Brake type | Inner shoe |
| Displacement | 11,867 cc | Foot brake action | On drive wheels |
| Compression ratio | 6.2-6.5:1 | Hand brake action | On drive wheels |
| RPM normal/maximum | 2600/3000 | Wheel type | Road wheels 520 x 95-397 |
| Maximum power | 265/300 HP | | Return rollers 310 x 70-203 |
| Power to weight | 13.6 HP/ton | Track | 2510 mm |
| Valves | Dropped | Track length | 2860 mm, 93 links |
| Crankshaft | 7 roller bearings | Track width | 400 mm |
| Carburetors | 2 Solex 40 JFF II | Ground clearance | 375 mm |
| Firing order | 1-12-5-8-3-10-6-7-2-11-4-9 | Overall length | 5400 mm |
| Starter | Bosch BNG 4/24 ARS 129 + Bosch AL/ZMD/R 9 | Overall width | 2950 mm |
| | | Overall height | 1960 mm |
| Generator | Bosch GTLN 600/12-1500 | Ground pressure | 0.93 kp/sq. cm |
| Batteries | 2 12-volt 105-Ah | Chassis weight | 15,800 kp |
| Fuel pumps | 2 mechanical + 1 electric | Allowable gross weight | 22,000 kp |
| Cooling | Water | Load limit | 1500 kp |
| Clutch | F & S dry 3-plate La 120 HDA | Seats | 4 |
| Gearbox | ZF SSG 77 Aphon | Fuel consumption | 195 liters/100 km |
| Speeds | 6 forward, 1 reverse | Oil consumption | Variable |
| Drive wheels | Front | Fuel capacity | 310 liters |
| Drive axle ratio | 1:4 (side rods) | Armor: hull front | 50 mm |
| Top speed | 40 kph | sides + rear | 30 mm |
| Range | Road 165, off-road 95 km | Performance: | |
| Steering | Daimler-Benz/Wilson clutch-type | Climbing ability | 30 degrees |
| Turning circle | 5.85 meters | Climbing distance | 600 mm |
| Track type | Kgs 61/400/120 | Wading | 800 mm |
| Suspension | 12 transverse torsion bars | Spanning | 2590 mm |
| Lubrication | High-pressure | Armament | 1 75mm StuK L.24 (44) |
| | | Intended use | Infantry escort gun |

## Gepanzerte Selbstfahrlafette für Sturmgeschütz 75mm Sturmkanone (Sd.Kfz.142) Ausf. E "7/ZW"

Manufacturer: Altmärkische Kettenwerk GmbH, Berlin-Spandau, Falkenhausen and other works
Year built: 1941

Source of information: D 652/43a of April 1, 1943
Chassis numbers: 90 751-91 034

| | |
|---|---|
| Engine make & type | Maybach "HL 120 TRM" |
| Cylinders | 60-degree V-12 |
| Bore x stroke | 105 x 115 mm |
| Displacement | 11,867 cc |
| Compression ratio | 6.2-6.5:1 |
| RPM normal/maximum | 2600/3000 |
| Maximum power | 265/300 HP |
| Power to weight | 13.5 HP/ton |
| Valves | Dropped |
| Crankshaft | 7 roller bearings |
| Carburetors | 2 Solex 40 JFF II |
| Firing order | 1-12-6-8-3-10-6-7-2-11-4-9 |
| Starter | Bosch BNG 4/24 ARS 129 + Bosch AL/ZMD/R 9 |
| Generator | Bosch GTLN 600/12-1500 |
| Batteries | 2 12-volt 105-Ah |
| Fuel pumps | 2 mechanical + 1 electric |
| Cooling | Water |
| Clutch | F & S dry 3-plate La 120 HDA |
| Gearbox | ZF SSG 77 Aphon |
| Speeds | 6 forward, 1 reverse |
| Drive wheels | Front |
| Drive axle ratio | 1:4 (side rods) |
| Top speed | 40 kph |
| Range | Road 165, off-road 95 km |
| Steering | Daimler-Benz/Wilson clutch-type |
| Turning circle | 5.85 meters |
| Suspension | 12 transverse torsion bars |
| Lubrication | High-pressure |
| Brake system | Daimler-Benz hydraulic, with servo |
| Brake type | Inner shoe |
| Foot brake action | On drive wheels |
| Hand brake action | On drive wheels |
| Wheel type | Road wheels 520 x 95-397 Return rollers 310 x 70-203 |
| Track | 2510 mm |
| Track length | 2860 mm, 93 links |
| Track width | 400 mm |
| Ground clearance | 375 mm |
| Overall length | 5400 mm |
| Overall width | 2950 mm |
| Overall height | 1960 mm |
| Ground pressure | 0.95 kp/sq. cm |
| Chassis weight | 15,800 kp |
| Allowable gross weight | 22,200 kp |
| Load limit | 1500 kp |
| Seats | 4 |
| Fuel consumption | 195 liters/100 km |
| Oil consumption | Variable |
| Fuel capacity | 310 liters |
| Armor: hull front | 50 mm |
| sides + rear | 30 mm |
| Performance: | |
| Climbing ability | 30 degrees |
| Climbing distance | 600 mm |
| Wading | 800 mm |
| Spanning | 2590 mm |
| Armament | 1 75mm StuK L/24 (50) |
| Intended use | Infantry escort gun |

**75mm Sturmgeschütz 40 (Sd.Kfz.142)**
**Ausf. F "7/ZW"**

Manufacturer: Altmärkische Kettenwerk GmbH, Falkensee, Spandau (assembly) and other works
Years built: 1941-42

Source of information: D 652/41 of April 1, 1943
Chassis numbers: 91035-91400

| | | | |
|---|---|---|---|
| Engine make & type | Maybach "HL 120 TRM" | Brake system | Daimler-Benz hydraulic, with servo |
| Cylinders | 60-degree V-12 | | |
| Bore x stroke | 105 x 115 mm | Brake type | Inner shoe |
| Displacement | 11,867 cc | Foot brake action | On drive wheels |
| Compression ratio | 6.2-6.5:1 | Hand brake action | On drive wheels |
| RPM normal/maximum | 2600/3000 | Wheel type | Road wheels 520 x 95-397 |
| Maximum power | 265/300 HP | | Return rollers 310 x 70-203 |
| Power to weight | 12.9 HP/ton | Track | 2510 mm |
| Valves | dropped | Track length | 2860 mm, 93 links |
| Crankshaft | 7 roller bearings | Track width | 400 mm |
| Carburetors | 2 Solex 40 JFF II | Ground clearance | 390 mm |
| Firing order | 1-12-5-8-3-10-6-7-2-11-4-9 | Overall length | 6770 +/5400 - gun |
| Starter | Bosch BNG 4/24 ARS 129 | Overall width | 2950 mm |
| | + Bosch AL/ZMD/R 9 | Overall height | 2150 mm |
| Generator | Bosch GTLN 600/12-1500 | Firing height | 1550 mm |
| Batteries | 2 12-volt 105-Ah | Ground pressure | 1.10 kp/sq. cm |
| Fuel pumps | 2 mechanical + 1 electric | Chassis weight | 15,800 kp |
| Cooling | Water | Allowable gross weight | 23,200 kp |
| Clutch | F & S dry 3-plate La 120 HDA | Load limit | 1500 kp |
| Gearbox | ZF SSG 77 Aphon | Seats | 4 |
| Speeds | 6 forward, 1 reverse | Fuel consumption | 182 liters/100 km |
| Drive wheels | Front | Oil consumption | Variable |
| Drive axle ratio | 1:4 (side rods) | Fuel capacity | 320 liters |
| Top speed | 40 kph | Armor: hull front | 50 mm |
| Range | Road 140, off-road 85 km |   sides & back | 30 mm |
| Steering | Daimler-Benz/Wilson clutch-type | Performance: | |
| Turning circle | 5.85 meters |   Climbing ability | 30 degrees |
| Track type | Kgs 61/400/120 |   Climbing distance | 600 mm |
| Suspension | 12 transverse torsion bars |   Wading | 900 mm |
| Lubrication | High-pressure |   Spanning | 2300 mm |
| | | Armament | 1 75mm StuK 40 L/43 (44) |
| | | | + 1 MG 34 (600) |
| | | Intended use | Infantry escort gun |

# 75mm Sturmgeschütz 40 (Sd.Kfz.142/1) Ausf. F/8 & G "8/ZW"

Manufacturer: Altmärkische Kettenwerk GmbH. Berlin-Spandau (assembly) and other works
Years built: 1942-45

Source of information: D 652/41 of April 1, 1943
Chassis numbers: Ausf. F/8 91401-91650, Ausf. G 91651-15001

| | | | |
|---|---|---|---|
| Engine make & type | Maybach "HL 120 TRM" | Brake type | Inner shoe |
| Cylinders | 60-degree V-12 | Foot brake action | On drive wheels |
| Bore x stroke | 105 x 115 mm | Hand brake action | On drive wheels |
| Displacement | 11,867 cc | Wheel type | Road wheels 520 x 95-397 |
| Compression | 6.2-6.5:1 | | Return rollers 310 x 70-203 |
| RPM normal/maximum | 2600/3000 | Track | 2510 mm |
| Maximum power | 265/300 HP | Track length | 2860 mm, 93 links |
| Power to weight | 12.5 HP/ton | Track width | 400 mm |
| Valves | Dropped | Ground clearance | 390 mm |
| Crankshaft | 7 roller bearings | Overall length | 6700 mm |
| Carburetors | 2 Solex 40 JFF II | Overall width | 2950/3330 with eastern tracks, |
| Firing order | 1-12-5-8-3-10-6-7-2-11-4-9 | | 3410 with aprons |
| Starter | Bosch BNG 4/24 ARS 129 | Overall height | 2160 mm |
| | + Bosch AL/ZMD/R 9 | Firing height | 1570 mm |
| Generator | Bosch GTLN 600/12-1500 | Ground pressure | 1.04 kp/sq. cm |
| Batteries | 2 12-volt 105-Ah | Chassis weight | 15,800 kp |
| Fuel pumps | 2 mechanical + 1 electric | Allowable gross weight | 23,900 kp |
| Cooling | Water | Load limit | 2000 kp |
| Clutch | F & S dry 3-plate La 120 HDA | Seats | 4 |
| Gearbox | ZF SSG 77 Aphon | Fuel consumption | Road 200, off-road 100 liters/km |
| Speeds | 6 forward, 1 reverse | Oil consumption | Variable |
| Drive wheels | Front | Fuel capacity | 310 liters |
| Drive axle ratio | 1:4 (side bars) | Armor: hull front | 50 + 30 mm |
| Top speed | 40 kph | sides & rear | 30 mm |
| Range | Road 155, off-road 95 km | Performance: | |
| Steering | Daimler-Benz/Wilson clutch-type | Climbing ability | 30 degrees |
| Turning circle | 5.85 meters | Climbing distance | 600 mm |
| Track type | Kgs 61/400/120 | Wading | 800 mm |
| Suspension | 12 transverse torsion bars | Spanning | 2300 mm |
| Lubrication | High-pressure | Armament | 1 75mm StuK 40 L/48 (54) |
| Brake system | Daimler-Benz mechanical | | + 1 MG 34 (600) |
| | | Intended use | Panzerjäger vehicle |

# 105mm Sturmhaubitze 42
## (Sd.Kfz.142/2) Ausf. G "10/ZW"

Manufacturer: Altmärkische Kettenfabrik GmbH, Berlin-Spandau (assembly) and other works

Years built: 1942-44
Source of information: D 652/41a of May 1, 1943

---

| | | | |
|---|---|---|---|
| Engine make & type | Maybach "HL 120 TRM Type A" | Foot brake action | On drive wheels |
| Cylinders | 60-degree V-12 | Hand brake action | On drive wheels |
| Bore x stroke | 105 x 115 mm | Wheel type | Road wheels 520 x 95-397 |
| Displacement | 11,867 cc | | Return rollers 310 x 70-203 |
| Compression ratio | 6.2-6.5:1 | Track | 2510 mm |
| RPM normal/maximum | 2600/3000 | Track length | 2869 mm, 93 links |
| Maximum power | 265/300 HP | Track width | 400 mm |
| Power to weight | 12.5 HP/ton | Ground clearance | 390 mm |
| Valves | Dropped | Overall length | 6140 mm |
| Crankshaft | 7 roller bearings | Overall width | 2950/3330 mm with eastern tracks, 3410 with aprons |
| Carburetors | 2 Solex 40 JFF II | | |
| Firing order | 1-12-5-8-3-10-6-7-2-11-4-9 | Overall height | 2150 mm |
| Starter | Bosch BNG 4/24 ARS 129 + Bosch AL/ZMD/R 9 | Firing height | 1570 mm |
| | | Ground pressure | 1.15 kp/sq. cm |
| Generator | Bosch GTLN 600/12-1500 | Chassis weight | 15,800 kp |
| Batteries | 2 12-volt 105-Ah | Allowable gross weight | 23,900 kp |
| Fuel pumps | 2 mechanical + 1 electric | Load limit | 1500 kp |
| Cooling | Water | Seats | 4 |
| Clutch | F & S dry 3-plate La 120/HDA | Fuel consumption | Road 200, off-road 330 liters/100 km |
| Gearbox | ZF SSG 77 Aphon | | |
| Speeds | 6 forward, 1 reverse | Oil consumption | Variable |
| Drive wheels | Front | Fuel capacity | 310 liters |
| Drive axle ratio | 1:4 (side rods) | Armor: hull front | 50 + 30 mm |
| Top speed | 40 kph | sides + rear | 30 mm |
| Range | Road 150, off-road 95 km | Performance: | |
| Steering | Daimler-Benz/Wilson clutch-type | Climbing ability | 30 degrees |
| Turning circle | 5.85 meters | Climbing distance | 600 mm |
| Track type | Kgs 61/400/120 | Wading | 800 mm |
| Suspension | 12 transverse torsion bars | Spanning | 2300 mm |
| Lubrication | High-pressure | Armament | 1 105mm StuH 42 L/28 (36) + 1 MG 34 or 42 (600) |
| Brake system | Daimler-Benz mechanical | | |
| Brake type | Inner shoe | Intended use | Assault howitzer for direct infantry support |

# Panzerbefehlswagen III (Sd.Kfz.266/267/268) Ausf. D¹, Früher Ausf. A, also "Grosser Panzerbefehlswagen" "3c/ZW"

Manufacturer: Daimler-Benz AG, Berlin-Marienfelde works
Years built: 1938-40

Source of information: D 652/20 of January 20, 1939
Chassis numbers: 60 341-60 370

| | |
|---|---|
| Engine make & type | Maybach "HL 120 TR" |
| Cylinders | 60-degree V-12 |
| Bore x stroke | 105 x 115 mm |
| Displacement | 11,867 cc |
| Compression ratio | 6.2-6.5:1 |
| RPM normal/maximum | 2600/3000 |
| Maximum power | 265/300 HP |
| Power to weight | 15.5 HP/ton |
| Valves | Dropped |
| Crankshaft | 7 roller bearings |
| Carburetors | 2 Solex 40 JFF II |
| Firing order | 1-12-5-8-3-10-6-7-2-11-4-9 |
| Starter | Bosch BNG 4/24 ARS 129 + Bosch AL/ZMD/R 9 |
| Generator | Bosch GTLN 700/12-1500 |
| Batteries | 2 12-volt 105-Ah |
| Fuel pumps | No data |
| Cooling | Water |
| Clutch | Dry multi-plate |
| Gearbox | ZF SSG 76 Aphon |
| Speeds | 6 forward, 1 reverse |
| Drive wheels | Front |
| Drive axle ratio | |
| Top speed | 55 kph |
| Range | Road 65, off-road 95 km |
| Steering | Daimler-Benz/Wilson clutch-type |
| Turning circle | 5.85 meters |
| Suspension | 3 longitudinal leaf springs for 8-wheel running gear |
| Lubrication | High-pressure |
| Brake system | Daimler-Benz mechanical with hydraulic support |
| Brake type | Inner shoe |
| Foot brake action | On drive wheels |
| Hand brake action | On drive wheels |
| Wheel type | Road wheels + return rollers |
| Track | 2490 mm |
| Track length | 3200 mm |
| Track width | 360 mm |
| Ground clearance | 375 mm |
| Overall length | 6000 mm |
| Overall width | 2870 mm |
| Overall height | 2450 mm |
| Firing height | 1890 mm |
| Ground pressure | 0.95 kp/sq. cm |
| Chassis weight | 13,300 kp |
| Allowable gross weight | 19,300 kp, net 18,100 kp |
| Load limit | 1200 kp |
| Seats | 5 |
| Fuel consumption | Road 180, off-road 320 liters/100 km |
| Oil consumption | Variable |
| Fuel capacity | 300 liters |
| Armor: hull + turret | 300 mm all around |
| Performance: | |
| Climbing ability | 30 degrees |
| Climbing distance | 600 mm |
| Wading | 800 mm |
| Spanning | 2300 mm |
| Armament | 2 MG 34 (1500) |
| Intended use | Heavy armored command vehicle for Panzer units |

**Panzerbefehlswagen III (Sd.Kfz.266/267/268)**
**Ausf. E, Früher Ausf. B "4a/ZW"**

Manufacturer: Daimler-Benz AG, Berlin-Marienfelde works
Year built: 1940

Source of information: D 652/17 of April 23, 1940
Chassis numbers: 60 501-60 545

| | | | |
|---|---|---|---|
| Engine make & type | Maybach "HL 120 TRM" | Brake system | Daimler-Benz hydraulic, with servo |
| Cylinders | 60-degree V-12 | | |
| Bore x stroke | 105 x 115 mm | Brake type | Inner shoe |
| Displacement | 11,867 cc | Foot brake action | On drive wheels |
| Compression ratio | 6.2-6.5:1 | Hand brake action | On drive wheels |
| RPM normal/maximum | 2600/3000 | Wheel type | Road wheels 520 x 95-398 |
| Maximum power | 265/300 HP | | Return rollers 310 x 70-203 |
| Power to weight | 15.3 HP/ton | Track | 2490 mm |
| Valves | Dropped | Track length | 2860 mm |
| Crankshaft | 7 roller bearings | Track width | 360 mm |
| Carburetors | 2 Solex 40 JFF II | Ground clearance | 385 mm |
| Firing order | 1-12-5-8-3-10-6-7-2-11-4-9 | Overall length | 5380 mm |
| Starter | Bosch BNG 4/24 ARS 129 + Bosch AL/ZMD/R 9 | Overall width | 2910 mm |
| | | Overall height | 2435 mm |
| Generator | Bosch GTLN 700/12-1500 | Ground pressure | 0.99 kp/sq. cm |
| Batteries | 2 12-volt 105-Ah | Chassis weight | 13,800 kp |
| Fuel pumps | 2 mechanical | Allowable gross weight | 19,500 kp |
| Cooling | Water | Load limit | 2000 kp |
| Clutch | Hydraulic with accelerator | Seats | 5 |
| Gearbox | Maybach SRG 32 8 145 Variorex pre-selector | Fuel consumption | 187 liters/100 km |
| | | Oil consumption | Variable |
| Speeds | 10 forward, 1 reverse | Fuel supply | 320 liters |
| Drive wheels | Front | Armor: hull + turret | 30 mm all around |
| Drive axle ratio | 1:4 (side rods) | Performance: | |
| Top speed | 40 kph | Climbing ability | 30 degrees |
| Range | Road 165, off-road 95 km | Climbing distance | 600 mm |
| Steering | Daimler-Benz/Wilson clutch-type | Wading | 800 mm |
| Turning circle | 5.85 meters | Spanning | 2300 mm |
| Suspension | 12 transverse torsion bars | Armament | 2 MG 34 |
| Lubrication | High-pressure | Intended use | Heavy command vehicle for Panzer units |

## Panzerbefehlswagen III (Sd. Kfz. 266/267/268) Ausf. H, Früher Ausf. C "7a/ZW"

Manufacturer: Daimler-Benz AG, Berlin-Marienfelde works
Years built: 1940-41

Source of information: D 652.17 of April 23, 1940
Chassis numbers: 70 001-70 145

| | | | |
|---|---|---|---|
| Engine make & type | Maybach "HL 120 TRM" | Brake type | Inner shoe |
| Cylinders | 60-degree V-12 | Foot brake action | On drive wheels |
| Bore x stroke | 105 x 115 mm | Hand brake action | On drive wheels |
| Displacement | 11,867 cc | Wheel type | Road wheels 520 x 95-397 |
| Compression ratio | 6.2-6.5:1 | | Return rollers 310 x 70-203 |
| RPM normal/maximum | 2600/3000 | Track | 2510 mm |
| Maximum power | 265/300 HP | Track length | 2860 mm, 93 links |
| Power to weight | 13.9 HP/ton | Track width | 400 mm |
| Valves | Dropped | Ground clearance | 385 mm |
| Crankshaft | 7 roller bearings | Overall length | 5520 mm |
| Carburetors | 2 Solex 40 JFF II | Overall width | 2950 mm |
| Firing order | 1-12-5-8-3-10-6-7-2-11-4-9 | Overall height | 2500 mm |
| Starter | Bosch BNG 4/24 ARS 129 + Bosch AL/ZMD/R 9 | Ground pressure | 0.93 kp/sq. cm |
| | | Chassis weight | 15,800 kp |
| Generator | Bosch GTLN 600/12-1500 | Allowable gross weight | 21,600 kp |
| Battery | 2 12-volt 105-Ah | Load limit | 2000 kp |
| Fuel pumps | 2 mechanical + 1 electric | Seats | 5 |
| Cooling | Water | Fuel consumption | 182 liters/100 km |
| Clutch | F & S dry 3-plate La 120 HDA | Oil consumption | Variable |
| Gearbox | ZF SSG 77 Aphon | Fuel capacity | 320 liters |
| Speeds | 6 forward, 1 reverse | Armor: hull front | 30 + 30 mm |
| Drive wheels | Front | sides + rear | 30 mm |
| Drive axle ratio | 1:4 (side rods) | turret | 30 mm all around |
| Top speed | 40 kph | Performance: | |
| Range | Road 165, off-road 95 km | Climbing ability | 30 degrees |
| Steering | Daimler-Benz/Wilson clutch-type | Climbing distance | 600 mm |
| | | Wading | 800 mm |
| Turning circle | 5.85 meters | Spanning | 2590 mm |
| Track type | Kgs 61/400/120 | Armament | 2 MG 34 |
| Suspension | 12 transverse torsion bars | Intended use | Heavy command vehicle for Panzer units |
| Lubrication | High-pressure | | |
| Brake system | Daimler-Benz hydraulic with servo | | |

## Panzerbefehlswagen III (Sd.Kfz.266/267/268) Ausf. K "8a/ZW"

Manufacturer: Daimler-Benz AG, Berlin-Marienfelde works
Years built: 1942-43
Source of information: D 652/26 of November 1, 1942

Sd.Kfz.266: Fu 6 & Fu 2
Sd.Kfz.267: Fu 6 & Fu 8
Sd.Kfz.268: Fu 6 & Fu 7
Chassis numbers: 70 201-70 400

| | | | |
|---|---|---|---|
| Engine make & type | Maybach "HL 120 TRM Type A" | Foot brake action | On drive wheels |
| Cylinders | 60-degree V-12 | Hand brake action | On drive wheels |
| Bore x stroke | 105 x 115 mm | Wheel type | Road wheels 520 x 95-397 |
| Displacement | 11,867 cc | | Return rollers 310 x 70-203 |
| Compression ratio | 6.2-6.5:1 | Track | 2510 mm |
| RPM normal/maximum | 2600/3000 | Track length | 2860 mm |
| Maximum power | 265/300 HP | Track width | 400 mm |
| Power to weight | 13.0 HP.ton | Ground clearance | 385 mm |
| Valves | Dropped | Overall length | 5520 mm |
| Crankshaft | 7 roller bearings | Overall width | 2950 mm |
| Carburetors | 2 Solex 40 JFF II | Overall height | 2500 mm |
| Firing order | 1-12-5-8-3-10-6-7-2-11-4-9 | Firing height | 1900 mm |
| Starter | Bosch BNG 4/24 ARS 129 + Bosch AL/ZMD/R 9 | Ground pressure | 1 kp/sq. cm |
| | | Chassis weight | 15,800 kp |
| Generator | Bosch GTLN 600/12-1500 | Allowable gross weight | 23,000 kp |
| Batteries | 2 12-volt 105-Ah | Load limit | 1500 kp |
| Fuel pumps | 2 mechanical + 1 electric | Seats | 5 |
| Cooling | Water | Fuel consumption | 160 liters/100 km |
| Clutch | F & S dry 3-plate La 120 HDA | Fuel consumption | Variable |
| Gearbox | ZF SSG 77 Aphon | Armor: hull front | 50 mm |
| Speeds | 6 forward, 1 reverse | sides | 30 mm |
| Drive wheels | Front | rear | 50 mm |
| Drive axle ratio | 1:4 (side rods) | turret front | 50 mm |
| Top speed | 40 kph | sides + rear | 30 mm |
| Range | Road 155, off-road 95 km | Performance: | |
| Steering | Daimler-Benz/Wilson clutch-type | Climbing ability | 50 mm |
| Turning circle | 5.85 meters | Climbing distance | 600 mm |
| Track type | Kgs 61/400/120 | Wading | 800 mm |
| Suspension | 12 transverse torsion bars | Spanning | 2200 mm |
| Lubrication | High-pressure | Armament | 1 50mm KwK 39 L/60 + 1 MG 34 (1500) |
| Brake system | Daimler-Benz mechanical | | |
| Brake type | Inner shoe | Intended use | Command vehicle for Panzer almost identical to tank. Gunner is also second radioman. |

## Panzerbeobachtungswagen III (Sd.Kfz.143) "ZW 38"

Manufacturer: Altmärkische Kettenwerk GmbH, Berlin-Spandau
Years built: 1942/43

Source of information: D 652.27 of April 23, 1940 and others

| | | | |
|---|---|---|---|
| Engine make & type | Maybach "HL 120 TRM" | Foot brake action | On drive wheels |
| Cylinders | 60-degree V-12 | Hand brake action | On drive wheels |
| Bore x stroke | 105 x 115 mm | Wheel type | Road wheels 520/95-398 |
| Displacement | 11,867 cc | | Return rollers 310/70-203 |
| Compression ratio | 6.2-6.5:1 | Track | 2510 mm |
| RPM normal/maximum | 2600/3000 | Track length | 2860 mm, type Kgs 61/400/120 |
| Maximum power | 265/300 HP | Track width | 400 mm, 93 links |
| Power to weight | 13.0 PS/ton | Ground clearance | 385 mm |
| Valves | Dropped | Overall length | 5520 mm |
| Crankshaft | 7 roller bearings | Overall width | 2950 mm |
| Carburetors | 2 Solex 40 JFF II | Overall height | 2500 mm |
| Firing order | 1-12-5-8-3-10-6-7-2-11-4-9 | Ground pressure | 1.01 kp/sq. cm |
| | + Bosch AL/ZMD/R 9 | Chassis weight | 15,800 kp |
| Generator | Bosch GTLN 600/12-1500 | Allowable gross weight | 23,000 kp |
| Batteries | 2 12-volt 105-Ah | Turret weight | 2500 kp |
| Fuel pumps | 2 Solex mechanical + 1 electric | Load limit | 1500 kp |
| Cooling | Water | Seats | 5 |
| Clutch | F & S dry 3-plate La 120/HDA | Fuel consumption | 182 liters.100 km |
| Gearbox | ZF SSG 77 Aphon | Oil consumption | Variable |
| Speeds | 6 forward, 1 reverse | Fuel capacity | 320 liters |
| Drive wheels | Front | Armor: hull front | 50 + 20 mm |
| Drive axle ratio | 1:4 (side rods) | sides | 30 mm |
| Top speed | Road 40/off-road 20 kph | rear | 50 mm |
| Range | Road 155, off-road 95 km | turret front | 50 mm |
| Steering | Daimler-Benz/Wilson clutch-type | sides + rear | 30 mm |
| Turning circle | 5.85 meters | Performance: | |
| Suspension | 12 transverse torsion bars | Climbing ability | 30 degrees |
| Lubrication | High-pressure | Climbing distance | 600 mm |
| Brake system | Daimler-Benz mechanical | Wading | 800 mm |
| Brake type | Inner shoe | Spanning | 2000 mm |
| | | Armament | 1 MG 34 |
| | | Intended use | Armored observation vehicle for armored artillery units |

## Sturminfanteriegeschütz 33

Manufacturer: Altmärkische Kettenwerk GmbH, Berlin-Spandau works (assembly)

Years built: 1941-42
Source of information: D 652/26 of November 1, 1942

| | | | |
|---|---|---|---|
| Engine make & type | Maybach "HL 120 TR, Type A" | Hand brake action | On drive wheels |
| Cylinders | 60-degree V-12 | Wheel type | Road wheels 520 x 95-397 |
| Bore x stroke | 105 x 115 mm | | Return rollers 310 x 70-203 |
| Displacement | 11,867 cc | Track | 2510 mm |
| Compression ratio | 6.2-6.5:1 | Track length | 2860 mm |
| RPM normal/maximum | 2600/3000 | Track width | 400 mm |
| Maximum power | 265/300 HP | Ground clearance | 400 mm |
| Power to weight | 14.3 HP/ton | Overall length | 5400 mm |
| Valves | Dropped | Overall width | 2900 mm |
| Crankshaft | 7 roller bearings | Overall height | 2300 mm |
| Carburetors | 2 Solex 40 JFF II | Firing height | 1905 mm |
| Firing order | 1-12-5-8-3-10-6-7-2-11-4-9 | Ground pressure | 0.89 kp/sq. cm |
| Starter | Bosch BNG 4/24 ARF 129 | Chassis weight | 15,800 kp |
| | + Bosch AL/ZMD/R 9 | Allowable gross weight | 21,000 kp |
| Generator | Bosch GTLN 600/12-1500 | Load limit | 1500 kp |
| Batteries | 2 12-volt 105-Ah | Seats | 5 |
| Fuel pumps | 2 mechanical + 1 electric | Fuel consumption | 300 liters/100 km |
| Cooling | Water | Oil consumption | Variable |
| Clutch | F & S dry 3-plate La 120/HDA | Fuel capacity | 320 liters |
| Gearbox | ZF SSG 77 Aphon | Armor: hull front | 50 mm |
| Speeds | 6 forward, 1 reverse | sides + rear | 30 mm |
| Drive wheels | Front | upper front | 80 mm |
| Drive axle ratio | 1:4 (side rods) | upper sides | 50 mm |
| Top speed | 20 kph | upper rear | 15 mm |
| Range | Road 110, off-road 85 km | Performance: | |
| Steering | Daimler-Benz/Wilson clutch-type | Climbing ability | 30 degrees |
| Turning circle | 5.85 meters | Climbing distance | 600 mm |
| Suspension | 12 transverse torsion bars | Wading | 800 mm |
| Lubrication | High-pressure | Spanning | 2300 mm |
| Brake system | Daimler-Benz mechanical | Armament | 1 150mm sIG 33 L/11 (30) + 1 MG 34 |
| Brake type | Inner shoe | | |
| Foot brake action | On drive wheels | Intended use | Test vehicle for armored mounting of sIG 33 |

## Panzerkampfwagen VK. 2001 (DB)

Manufacturer: Daimler-Benz AG, Berlin-Marienfelde works
Years built: 1939-40

Source of information: Daimler-Benz AG archives
2 prototypes built

| | | | |
|---|---|---|---|
| Engine make & type | Daimler-Benz MB 809 | Lubrication | Central |
| Cylinders | 60-degree V-12 | Brake system | Daimler-Benz mechanical with hydraulic activation |
| Bore x stroke | 115 x 140 mm | | |
| Displacement | 17,500 cc | Brake type | Servo inner shoes |
| Compression ratio | 14.5:1 | Foot brake action | On drive wheels |
| RPM | 2400 | Hand brake action | On drive wheels |
| Maximum power | 350-360 | Wheel type | Road wheels + return rollers |
| Power to weight | 16.3 HP/ton | Track | |
| Valves | Dropped | Track length | 2757 mm |
| Crankshaft | 7 journal bearings | Track width | 440 mm |
| Fuel injection pump | 1 Bosch PE 12 | Ground clearance | 425 mm |
| Firing order | 1-8-5-10-3-7-6-11-2-9-4-12 | Overall length | 5130 mm |
| Starter | Bosch 24 V + compressed air | Overall width | 3020 mm |
| Generator | Bosch 12-volt, 600-watt | Overall height | 1640 mm minus turret |
| Batteries | 2 12-volt 150-Ah | Ground pressure | |
| Fuel pumps | Unspecified | Chassis weight | |
| Cooling | Water | Allowable gross weight | 22,000 kp |
| Clutch | Multi-plate | Load limit | 1500 kp |
| Gearbox | Daimler-Benz/Wilson pre-selector* | Seats | 5 |
| | | Fuel consumption | 190 g/HP/h |
| Speeds | 8 forward, 1 reverse | Oil consumption | Variable |
| Drive wheels | Front | Fuel capacity | |
| Drive axle ratio | | Intended use | Prototype without turret, meant to replace Panzer III |
| Top speed | 50 kph | | |
| Range | | | |
| Steering | 3- or 4-radius overlapping gears** | Notes:<br>* Optional ZF 8-speed pre-selector gearbox, or<br>** DB clutch-type steering gear of the usual type. | |
| Turning circle | 6.0 meters | | |
| Suspension | Longitudinal leaf springs, one each for 4 front and 3 rear wheels | | |

# Bibliography

Willi A. Boelcke, *Deutschlands Rüstung im Zweiten Weltkrieg*
Uwe Feist, *Panzer III in action*
Heinz Guderian, *Erinnerungen eines Soldaten*
Fritz Heigl, *Taschenbuch der Tanks*
Robert J. Icks, *Tanks and Armored Vehicles*
P. Kantakoski, *Suomalaiset panssarivanujoukot 1919-1969*
Janusz Magnuski, *Wozy Bojowe*
F.W. von Mellenthin, *Panzer Battles*
Oskar Munzel, *Die deutschen gepanzerten Truppen bis 1945*
Walther K. Nehring, *Die Geschichte der deutschen Panzerwaffe 1916-1945*
R.M. Ogorkiewicz, *Armor*
Werner Oswald, *Kraftfahrzeuge und Panzer der Reichswehr, Wehrmacht und Bundeswehr*, Motorbuch-Verlag, Stuttgart
H. Scheibert-C. Wagener, *Die deutsche Panzertruppe 1939-1945*
F.M. von Senger und Etterlin, *Die deutschen Panzer 1926-1945*
Walter J. Spielberger-Friedrich Wiener, *Die deutschen Panzerkampfwagen III & IV*
Walter J. Spielberger, *Der Panzerkampfwagen III und seine Abarten 1935-1945*
Walter J. Spielberger, *Panzerkampfwagen III*
Walter J. Spielberger-Uwe Feist, Armor Series 1-10
Rolf Stoves, *Die 1. Panzer Division*
G. Tornau-F. Kurowski, *Sturmartillerie*

## Explanation of Abbreviations used

| | |
|---|---|
| A (2) | Infantry Department of the War Ministry |
| A (4) | Field Artillery Department of the War Ministry |
| A (5) | Foot Artillery Department of the War Ministry |
| A 7 V | Transportation Department of the War Ministry |
| AD (2) | War Department, Section 2 (Infantry) |
| AD (4) | War Department. Section 4 (Field Artillery) |
| AD (5) | War Department, Section 5 (Foot Artillery) |
| AHA/Ag K | Army Office, Motor Vehicle Group |
| AK | Artillery Design Bureau |
| AKK | Army Motor Vehicle Column |
| AlkW | Army Trucks |
| ALZ | Army Freight Unit |
| AOK | Army High Command |
| APK | Artillery Testing Commission |
| ARW | Eight-Wheeled Vehicle |
| A-Types | Vehicles with all-wheel drive (fast type) |
| Ausf | Ausfürung: model, type |
| BAK | Anti-Balloon Gun |
| Bekraft | Fuel Department, Field Motor Vehicles Department |
| Chefkraft | Chief of the Field Motor Vehicles Department |
| (DB) | Daimler-Benz |
| DMG | Daimler Motor Company |
| Dtschr. Krprz. | German Crown Prince |
| E-Fahrgestell | Standard chassis |
| E-Pkw | Uniform Passenger Vehicle |
| E-Lkw | Uniform Freight Vehicle |
| FA | Field Artillery |
| FAMO | Fahrzeug- und Motorenbau GmbH |
| FF-Kabel | Field Telephone Cable |
| FH | Field Howitzer |
| FK | Field Gun |
| Flak | Anti-Aircraft Gun |
| F.T. | Radio/Telegraph |
| Fu | Radio |
| Fu Ger | Radio device |
| Fr Spr Ger | Radio speaking device |
| g | Secret |
| Gen.St.d.H. | Army General Staff |
| Gengas | Generator Gas |
| G.I.d.MV. | General Inspection of Military Vehicles |
| g. Kdos | Secret command matter |
| gp | Armored |
| g. RS | Secret government matter |
| gl | Off-road capable |
| GPK | Gun Testing Commission |
| (H) | Rear engine |
| Hanomag | Hannoversche Maschinenbau AG |

| | | | |
|---|---|---|---|
| Hk | Halftrack | O. H. L. | Army Highest Command |
| H.Techn.V.Bl. | Army Technical Instruction Sheet | O. K. H. | Army High Command |
| HWA | Army Weapons Office | O. K. W. | Wehrmacht High Command |
| I. D. | Infantry Division | Pak | Antitank Gun, armor-piercing gun |
| I. G. | Infantry Gun | P. D. | Panzer Division, Armored Division |
| In. | Inspection | Pf | Engineer Vehicle |
| In. 6 | Inspection of Motor Vehicles | Pkw | Passenger Vehicle |
| Ikraft | Inspection of Field Motor Vehicles | Pz. F. | Armored Vehicle |
| Iluk | Inspection of Air and Motor Vehicles | Pz.Kpfwg | Battle Tank |
| K | Cannon, gun | Pz. Spwg. | Armored Scout Car |
| KD | Krupp-Daimler | (R) | Tracks |
| K. D. | Cavalry Division | R/R | Wheel with Track |
| KdF | Kraft durch Freude (Nazi organization) | (RhB) | Rheinmetall-Borsig |
| K.d.K. | Commander of Motor Vehicle Troops | RS | Track-type Tractor |
| K. Flak | Motorized Anti-Aircraft Gun | RSG | Mountain Track-type Tractor |
| Kfz. | Motor Vehicle | RSO | Tracked Vehicle East (Wheeled Tractor East) |
| KM | War Ministry | RV | Targeting Communications |
| KP | Motorized Limber | s | Heavy |
| (Kp) | Krupp | schg. | Running on rails |
| Kogenluft | Commanding General of the Air Combat Forces | schf. | Amphibious |
| Krad | Motorcycle | Sd. Kfz. | Special Motor Vehicle |
| Kr. Zgm. | Motor Tractor | Sfl. | Self-Propelled Gun |
| KS | Fuel Injection | Sf | Self-Propelled |
| Kw | Motor Vehicle, also Fighting Vehicle | S-Typen | Rear-wheel drive (fast types) |
| l | Light | SmK | Pointed bullet with core |
| L/ | Caliber length | SSW-Zug | Siemens-Schuckert Works Train |
| le | Light | s. W. S. | Heavy Wehrmacht Tractor |
| le FH | Light Field Howitzer | Tak | Antitank Gun |
| le FK | Light Field Gun | Takraft | Technical Dept, Inspection of Motor Vehicles |
| l. F. H. | Light Field Howitzer | TF | Carrier Frequency (radio) |
| le. I. G. | Light Infantry Gun | Tp | Tropical version |
| le. W. S. | Light Military Tractor | Vakraft | Test Department, Field Motor Vehicles (WWI) |
| LHB | Linke-Hofmann-Busch | Vkraft | Test Department, Inspection of Motor Vehicles (Reichswehr and Wehrmacht) |
| l. I. G. | Light Infantry Gun | | |
| Lkw | Freight Vehicle, Truck | ve | Fully suppressed car ignition |
| LWS | Land-Water Tractor | v/max | Top speed |
| m | Medium | Vo | Muzzle velocity |
| MAN | Maschinenfabrik Augsburg-Nürnberg AG | VK | Tracked Test Vehicle |
| MG | Machine Gun | VPK | Vehicle Technical Testing Commission |
| MP | Machine Pistol | Vs. Kfz. | Test Vehicle |
| MTW | Personnel Transport Vehicle | ZF | Zahnradfabrik Friedrichshafen |
| n = | Revolutions per minute | ZRW | Ten-Wheeled Vehicle |
| n/A | New type, new version | WaPruef/WaPrw | Weapon Testing Office |
| NAG | Nationale Automobilgesellschaft | Wumba | Weapon and Ammunition Procurement Office |
| (o) | Trade version | wg | Amphibious |
| Ob. d. H. | Commander of the Army | | |